The Importance of Intelligent Automation to Any Digital Transformation Strategy, How IA is a Critical Lynchpin to Digital Transformation Success

by

Jonathan T. Hardy

Copyright © 2024

All rights reserved. No part of this publication may be reproduced, distributed, or transmitted in any form or by any means, including photocopying, recording, or other electronic or mechanical methods, without the prior written permission of the publisher, except in the case of brief quotations embodied in critical reviews and certain other noncommercial uses permitted by copyright law.

Copyright, publication, and cover design by:

Jonathan T. Hardy

ISBN: 979-8-9914378-1-3

This book is dedicated:

To my beautiful and ever-supportive wife, Lateisha, I will love you until time's end, my God's gift. You have always believed I was worthy, even during periods I did not.

To my wonderful children, Morgan and Dylan, remember that I will always love and be proud of you.

To my father, Joseph Hardy, I know you would be so proud if you were still with us. Thank you for the man I have become.

To my mother, Wanda Hardy, thank you, Mom, for all the love and support you have given me. I will cherish you forever.

To my stepmother, Sharon Hardy, thank you for being the best "Mimi" a step-son could have ever had and for your ever-present love and devotion.

To my dear brothers and sisters and their spouses, I am the eldest and, by default, the coolest of us all to absolute zero. I love you all dearly.

To my friends, mentors, and colleagues over the years. I have learned so much from you and hope this book and its approach reflect well on all I have gained from you.

"Intelligent Automation is a central catalyst for Digital Transformation, driving efficiency, innovation, and strategic growth across industries. Our vision is to leverage automation to free human talent for higher-value work. By blending technology with human know-how, we aim to accelerate the transition to agile, future-ready, Digitally Native enterprises. Intelligent Automation is not just about operational excellence; it's about shaping a sustainable competitive edge in this emerging digital era with precision and professionalism."

Contents

PREFACE I

INTRODUCTION 1

PART 1: THE STRATEGIC IMPERATIVE OF
INTELLIGENT AUTOMATION 28

Chapter 1: Intelligent Automation as a Driver of
Business Model Innovation 29

Chapter 2: Transforming the Customer Experience
with IA 46

Chapter 3: IA as an Employee Empowerment Tool
.... 58

PART 2: BUILDING A FOUNDATION FOR DIGITAL
TRANSFORMATION SUCCESS BASED ON IA 73

Chapter 4: Developing a Strategic Roadmap for
Intelligent Automation 74

Chapter 5: The Critical Role of an Intelligent
Automation Framework in Digital Transformation
.... 94

Chapter 6: Fostering an Enterprise-Wide Automation
Development Mindset 127

Chapter 7: How Intelligent Automation Can Be Key to
Driving Strategic Value 159

PART 3: THE IMPORTANCE OF INTELLIGENT
AUTOMATION IN CRAFTING A DIGITAL
TRANSFORMATION STRATEGY 179

Chapter 8: Aligning IA with Digital Transformation Principles   180

Chapter 9: Sustaining Being Digitally Native Amid Rapid Tech Disruptions   202

PART 4: THE QUICKLY EMERGING FUTURE FOR IA AND DIGITAL TRANSFORMATION AND CLOSING REMARKS   231

Chapter 10: The Emerging Technologies That Are Driving the Path Forward   232

Chapter 11: Closing Thoughts on IA's Importance to Digital Transformation   269

Chapter 12: How Generative AI Powered the Writing of this Book   280

GLOSSARY   303

REFERENCES   314

Preface

In an age where digital technology reshapes boundaries and redefines possibilities, the strategic importance of Intelligent Automation (IA) in ensuring Digital Transformation success has never been more pronounced. "The Importance of Intelligent Automation to Any Digital Transformation Strategy, How IA is a Critical Lynchpin to Digital Transformation Success" provides a deep dive into the role of IA not just as a complementary capability but as a fundamental pillar in the Digital Transformation strategies of forward-thinking organizations.

As businesses face an ever-accelerating pace of change, adopting a comprehensive digital strategy incorporating Intelligent Automation becomes critical. The digital landscape offers untapped opportunities for those ready to invest in and integrate sophisticated IA technologies. These tools are no longer about just reducing operational costs; they are also about creating innovative, robust solutions that deliver exceptional value that support the growth and efficiency ambitions of the enterprises they support.

This book is designed to move away from thinking about Intelligent Automation from a tactical lens to thinking about it in a more holistic paradigm, offering a pathway from conceptual understanding to strategic application. It aims to guide decision-makers through the complexities of Digital Transformation with a clear focus on the strategic deployment of IA technologies. Here, IA is positioned as a critical enabler and an essential component that supports and accelerates the achievement of broader objectives via Digital Transformation goals.

Written for executive leaders and IA practitioners alike, the discussion in this book is designed to equip you with the necessary knowledge to understand the strategic benefits of IA, integrate it with core business processes, and harness its potential to drive unprecedented growth and competitiveness. The book covers the critical aspects of IA necessary for a successful Digital Transformation, from the foundational concepts of Robotic Process Automation (RPA) to the advanced realms of cognitive automation incorporating AI and Machine Learning.

The content is meticulously structured to facilitate a linear yet comprehensive exploration of Intelligent Automation. It starts with an introduction to IA, understanding of its roots, and the evolution from basic RPA to advanced IA. It then guides readers through the strategic implementation of IA, aligning it with business strategies to enhance operational efficiency, customer service, and innovation. The book also explores future trends in IA advancements and their implications for businesses looking to maintain a competitive edge in a digitally competitive world, providing real-world applications and success stories from various industries.

As you navigate through the stories and strategies outlined in this book, the goal is to inform and inspire. This book aims to catalyze leaders and organizations to embrace IA as a transformative force by demystifying Intelligent Automation and illustrating its profound impact on digital strategies. The future is Digitally Native, and Intelligent Automation is the key to unlocking this future understanding. Leveraging it will determine the success of businesses in an increasingly competitive corporate landscape.

We invite you to embark on this exploratory journey through the pages of this book. Each chapter builds upon the last to provide a detailed, actionable strategy that aligns with your Digital Transformation goals. Herein lies the blueprint for a digitally transcendent future powered by the strategic application of Intelligent Automation. Welcome to the journey.

Introduction

Digital Transformation and Intelligent Automation as Existential Needs of the Enterprise

Given the pace of technology today, the necessity for businesses to embrace Digital Transformation has transcended mere advantage and become a survival imperative. This transformation is a holistic reimagining of business processes, customer interactions, and the essence of corporate strategy through digital innovation.

RPA is a transformative technology that automates rule-based and repetitive tasks by mimicking human interactions with digital systems. Its evolution from simple task automation to sophisticated, Intelligent Automation incorporating Artificial Intelligence and Machine Learning reflects a significant leap in how businesses optimize operations, enhance efficiency, and innovate service delivery. This historical overview will delve into the origins, development, and maturation of RPA, highlighting key figures, companies, and milestones that have shaped its trajectory.

A Brief History of Robotic Process Automation

The emergence of RPA as a distinct field is attributed to several key players who recognized the potential of automating routine tasks with software "robots." Blue Prism, founded in 2001 by Alastair Bathgate and David Moss, is often credited with pioneering the RPA industry. They introduced one of the first platforms that enabled businesses to automate clerical tasks without altering their existing IT infrastructure (IntellPaat, n.d.).

RPA can trace its roots back to the early days of business process and workflow automation in the late 1990s and early 2000s, with its inception often attributed to the automation of User Interface (UI) testing in the 1990s. This period saw a shift towards automating visual elements of interfaces to ensure they function correctly. This era marked the beginning of using automation to streamline processes, driven by the advent of Windows 95, which significantly expanded the computer user base from corporate and governmental organizations to home users, necessitating the development of UI testing as screen sizes and requirements became more diverse. "Robotic Process Automation" was coined to describe the automation of tasks traditionally performed by human operators using computer systems, such as data entry, transaction processing, and simple administrative tasks.

Automation Anywhere and UiPath, founded in 2003 and 2005, respectively, joined the fray, each bringing innovative solutions that made RPA more accessible and powerful. Mihir Shukla, CEO of Automation Anywhere, and Daniel Dines, CEO of UiPath, have been instrumental in driving the adoption of RPA technologies across various industries. These companies have played pivotal roles in shaping the RPA landscape, contributing significantly to its development, adoption, and growth. Over the years, RPA has seen significant advancements in technology, scalability, and reliability. Early adoption was primarily in industries with high volumes of repetitive tasks, such as banking, insurance, and telecommunications. However, its application has since expanded across sectors, including healthcare, retail, and public services.

The wide recognition and adoption of RPA by large-scale businesses occurred around 2012 amidst a backdrop of economic recovery and a growing realization of the need for Digital Transformation. This period saw RPA emerge as a powerful tool for reducing expenses and streamlining operations, with its adoption rapidly spreading across enterprises seeking to automate mission-critical tasks.

The Evolution Towards Intelligent Automation

Robotic Process Automation has undergone a transformative journey, evolving into the broader and more complex landscape of Intelligent Automation. While steeped in technological advancement, this evolution is fundamentally a story of how the pursuit of efficiency, scalability, and innovation has pushed the boundaries of automation beyond its initial conception.

The genesis of RPA was marked by the automation of repetitive, rule-based tasks that required little to no human judgment. This revolutionary capability allowed businesses to significantly reduce manual workload, minimize errors, and improve operational efficiency. RPA's initial appeal lies in its ability to mimic human actions by software robots (or "bots"), executing well-defined, structured tasks with unparalleled speed and accuracy. However, as the need to automate more complex business processes grew, the limitations of RPA's rule and GUI nature became apparent. There was a burgeoning need for an automation platform that could develop solutions that could either handle the volume of complex enterprise processes, the ambiguity of those processes, or both in ways the capabilities of traditional RPA could not.

Introduction

The transition from RPA to Intelligent Automation represents a paradigm shift, integrating cognitive technologies such as ML, NLP, and AI with traditional RPA and coupling it with far greater and more robust data integration capabilities, like APIs and the ability to query databases directly. This evolution was driven by the recognition that to truly transform business operations, automation is needed to move from performing tasks to optimizing them, as the additions of ML, NLP, and AI allow. Intelligent Automation, therefore, emerged as a solution that could interpret unstructured data, adapt to changing environments, and make informed decisions based on the specifications of the cognitive solution coupled with the bot. By incorporating AI and ML, IA now has a learning element to enhance their efficiency over time and provide businesses with previously unattainable performance and insights. This strategy is about automating individual tasks and reimagining business processes and models to leverage digital technologies to their utmost potential. It entails a cultural, organizational, and operational change, acknowledging that the value of automation lies not just in cost reduction but in enabling innovation, improving customer experiences, and fostering a digital-first mindset.

A significant aspect of this evolution has been the development of governance frameworks and Centers of Excellence (CoEs) that ensure the alignment of IA initiatives with business strategies and goals. These frameworks are critical in managing the complexities associated with deploying IA technologies, ensuring compliance, and maximizing the value derived from automation efforts. By establishing clear governance, businesses can navigate the challenges of integrating IA, such as managing change, ensuring data security, and maintaining operational integrity.

The evolution from RPA to Intelligent Automation marks a shift from automating tasks to automating decision-making and insights. This journey reflects the broader Digital Transformation efforts undertaken by businesses worldwide, driven by the need to stay competitive. Intelligent Automation stands at the forefront of this shift, offering the tools and capabilities needed to transform how businesses operate, innovate, and deliver value.

Intelligent Automation and the Covid-19 Pandemic

As the COVID-19 pandemic swept across the world, businesses encountered unprecedented challenges, prompting an urgent need for adaptability and resilience. The sudden shift to remote work and the disruption of traditional business processes catalyzed a significant transformation in operational strategies. Intelligent Automation became a linchpin for organizations striving to maintain continuity and efficiency during these turbulent times (Coombs, 2020).

Initially, many companies were at various stages of their Digital Transformation journeys, with IA being a key component of their automation efforts. IA's ability to automate rule-based, repetitive tasks offered a solid foundation. However, the complexities and dynamic nature of the challenges presented by the pandemic necessitated a more sophisticated approach. Intelligent Automation, with its cognitive capabilities, allowed companies to expand their process automation scope, providing companies with a far greater range of use cases to consider for automation at that point.

The adoption of IA during the pandemic was not straightforward. For many organizations, accelerating their automation initiatives under pressing circumstances involved tireless effort and swift strategic pivoting. Businesses that were already familiar with RPA and had burgeoning automation programs were better positioned. Yet, scaling these solutions to address the enterprise's broader and more complex needs amidst the crisis required significant investment in time, resources, and strategic foresight.

During this period, the primary focus of Intelligent Automation was on business process automation, particularly in domains directly impacting the continuity of operations and remote work enablement. This included automating customer service interactions through AI-powered chatbots, streamlining administrative tasks to support the workforce's transition to remote work, and ensuring that critical business functions could operate seamlessly in a digital-only environment. The agility and efficiency provided by IA were instrumental in allowing companies to quickly adapt their processes, ensuring not only the continuity of operations but also the retention of customer trust and satisfaction in a time of widespread uncertainty.

Moreover, the pandemic underscored the importance of having a strategic, holistic approach to automation, one that transcends piecemeal solutions to foster enterprise-wide resilience. Organizations that viewed automation through an Enterprise Automation Development Mindset found themselves at a distinct advantage. This mindset emphasizes the integration of automation technologies across the entire organization, aligning with overarching business objectives to drive innovation, efficiency, and agility.

The COVID-19 pandemic acted as a catalyst, accelerating the adoption of Intelligent Automation by businesses seeking to navigate the crisis. While the journey to implementing IA was marked by tireless effort and rapid adaptation, the results underscored the value of technology in ensuring business continuity and operational resilience.

Case Study: The Value of Foresight: An RPA Story

It was January 2019 when Global Financial took the forward-looking step of establishing a Robotic Process Automation Center of Excellence (RPA CoE). At the time, it seemed like an operational nicety - a way to boost efficiency by automating some of the many repetitive, manual tasks that underpinned the company's operations. Little did they know just how valuable that investment would prove to be.

At the beginning of the Covid-19 pandemic in early 2020, Global Financial was still in the relatively early stages of its RPA journey. The CoE had deployed a handful of software robots to automate processes like data entry, report generation, and transaction processing. But compared to many other firms, they were ahead of the curve.

As lockdowns took hold across the world, the company's offshore operations in Bangalore and Manila ground to a halt.

Suddenly, a significant portion of the manual work that formed the backbone of Global Financial's back-office operations could no longer be performed. For firms without RPA, it was an existential crisis. However, for Global Financial, it was the catalyst for accelerating their automation efforts at breakneck speed.

Working remotely, the RPA team worked around the clock, putting in long days and nights to rapidly build, test, and deploy a fleet of new software robots. Working closely and intensively with the various processes' SMEs, they worked hard to spin up automations in a matter of days rather than weeks it might have taken otherwise. Recognizing that time was of the essence, they prioritized the highest-impact, most critical processes - anything that could prevent service disruptions for their clients.

It wasn't seamless, and there were more than a few hiccups and hurdles to overcome. But a shared sense of purpose drove the team to persevere. Everyone from developers to process experts to infrastructure teams rallied together with a common goal - to keep the business running.

Within several weeks, over 50 critical processes had been automated, helping to fill the operational void left by the shuttered offshore offices. Loan applications, account openings, transactions, reconciliations - all were now being processed efficiently by Global Financial's digital workforce.

As the pandemic stretched for months, it became abundantly clear that RPA was more than a short-term fix. Instead, it unlocked a new world of operational flexibility, agility, and resilience. The ability to rapidly scale up digital operations would prove to be a foundational component of Global Financial's Digital Transformation roadmap going forward.

What had started as an opportunistic experiment in process optimization had revealed the true power of Intelligent Automation. As the firm emerged from the pandemic, the RPA CoE would only grow in strategic importance. Its mission was clear: to serve as the catalyst for enterprise-wide operational transformation, leveraging the latest AI and automation technologies to continually increase efficiency, adaptability, and customer service, which was all thanks to the foresight of an early investment in RPA.

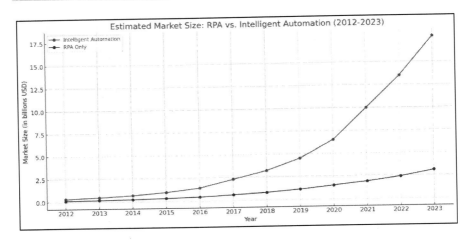

Figure Intro.1, (OpenAI, n.d.), The line chart includes the estimated market sizes for Intelligent Automation (IA) and Robotic Process Automation (RPA) from 2012 to 2023.

- Intelligent Automation shows a significantly higher and steeper growth trajectory compared to RPA alone, starting from $0.3 billion in 2012 and escalating to an estimated $17.8 billion by 2023. This reflects the robust adoption and integration of AI and Machine Learning technologies with automation solutions.
- RPA Only also exhibits growth but at a more gradual pace, increasing from $0.1 billion in 2012 to $3.1 billion in 2023. This trend indicates steady adoption within industries focused on automating routine, rule-based tasks.

The Imperative of Intelligent Automation's Role in Digital Transformation

In today's rapidly evolving digital scene, the imperative for organizations to undergo Digital Transformation has never been more important. This urgency stems from a complex interplay of technological advancements, changing consumer expectations, and the intensifying competitive pressure across industries. Digital Transformation, a strategic integration of digital technologies into all areas of a business, foundationally changes how value is delivered to customers by companies. It also fosters a culture of continuous innovation and adaptation, critical for achieving competitive advantage and ensuring long-term viability.

The transition from RPA to IA signifies a critical phase in an organization's Digital Transformation journey. It highlights the shift from automating tasks to reimagining entire business processes and models. It aligns with the broader Digital Transformation thought process, ensuring that automation initiatives are not siloed but a part of a comprehensive approach to leveraging digital technologies (Saldanha, 2019).

Digital Transformation, however, extends beyond the adoption of digital technologies. It requires a fundamental rethinking of business strategies, organizational structures, and culture. Organizations must embrace a digital-first mindset, where data inform decisions and technologies are leveraged to create new value propositions. This includes developing digital capabilities like data analytics, native application integration via APIs, and a cloud-based application infrastructure to support new business models and enhance customer experiences. Moreover, Digital Transformation is an ongoing process of change and adaptation. The rapid pace of change demands that organizations remain aware and responsive to emerging trends and innovations. This fosters a culture of learning and experimentation and encourages innovation. By doing so, organizations can stay ahead of the curve, anticipate customer needs, and respond to market changes with agility to become truly Digitally Native (Saldanha, 2019).

The urgency of Digital Transformation for competitive advantage and long-term viability cannot be overstated. In an increasingly digital world, organizations that fail to adapt to oncoming disruption risk becoming obsolete. Conversely, those who embrace Digital Transformation can unlock new opportunities for growth, efficiency, and customer engagement. They can navigate the complexities of today's digital age with confidence, secure in the knowledge that their strategies, processes, and cultures are aligned with the demands of a rapidly changing world.

Digital Transformation represents a critical pathway for organizations seeking to maintain relevance, achieve competitive advantage, and ensure long-term success. By integrating digital technologies like IA into their operations, reimagining their business processes, and fostering a culture of innovation, organizations can navigate the challenges and opportunities of the digital era. The journey is complex and continuous, but the rewards of growth, efficiency, and digital resilience are immense.

How Intelligent Automation Enables the Success of Digital Transformation

Intelligent Automation stands at the heart of modern Digital Transformation initiatives, a pivotal force propelling businesses from traditional operations into the future of streamlined, technology-driven processes. The journey of IA, particularly its evolution from basic Robotic Process Automation to more complex cognitive capabilities, marks a significant leap toward realizing an enterprise's digital potential. This narrative explores how IA fosters the success of Digital Transformation, weaving through its progression, integration, challenges, and future prospects within the broader digital landscape.

The inception of RPA laid the groundwork for IA. This foundational phase of automation focused on mimicking the manual actions of users within software applications, thus offering businesses their first taste of operational efficiency and accuracy improvements. However, the true essence of Digital

Transformation, which encompasses a comprehensive overhaul involving processes and cultural and strategic realignment, demanded more than what RPA could offer alone.

As businesses become more dependent on digital solutions for their competitive success, offering the means to automate processes and transform entire business models and operational frameworks has emerged as an ever-apparent reality. Given this truth, the strategic implementation of IA within Digital Transformation initiatives requires a holistic approach. Central to this strategy is the establishment of a governance framework for Intelligent Automation, ensuring that automation efforts align with the company's broader Digital Transformation goals and objectives. Such a framework provides the structure needed to manage, scale, and sustain automation initiatives, addressing key areas such as engagement, delivery, development, governance, infrastructure, and operations. This structured approach is essential in navigating the complexities of Digital Transformation, where the integration of new technologies must be balanced with organizational culture, customer expectations, and market dynamics.

A pivotal aspect of IA's role in Digital Transformation is its ability to enhance decision-making and operational efficiency. By leveraging data analytics via process mining, businesses can uncover insights that drive more informed strategic decisions, optimizing processes and delivering more tailored customer delivery. The shift to a data-driven organization is emblematic of this transformation, where IA not only automates processes but also enables the harnessing of data for strategic advantage. This transition is critical in an increasingly competitive landscape where agility, innovation, and being customer-focused are paramount. Moreover, adopting IA facilitates shifting toward more agile, resilient operational models. The COVID-19 pandemic from 2020 emphasized the need for businesses to adapt quickly to changing conditions, a challenge that IA is uniquely equipped to address.

However, the journey towards harnessing IA for Digital Transformation is fraught with challenges. Among the most significant are the cultural and organizational changes required to integrate and maximize the benefits of automation technologies. The successful deployment of IA demands a workforce that is not only skilled in new technologies but also adaptable to the changing nature of work. Furthermore, achieving alignment across different departments, ensuring data quality and security, and managing the ethical implications of automation are critical hurdles that organizations must overcome.

The journey of IA, from its roots in RPA to its role as a cornerstone of Digital Transformation, reflects the dynamic interplay between technology, strategy, and organizational culture. This view encapsulates the essence of the digital age, in which firms relentlessly pursue becoming Digitally Native in order to become more innovative, efficient, and creative to generate more value through the strategic application of technology.

The Importance of Defining a Digital Transformation Strategy

A Digital Transformation strategy, at its core, is a plan that outlines how an organization will leverage technology to transform its operations, culture, and customer interactions (Saldanha, 2019). This strategy is not just about adopting new technologies but about rethinking and radically changing how a business operates and delivers value to its customers. This strategic planning is critical because it aligns technological capabilities with the company's business objectives, ensuring that every tech investment contributes directly to its overarching goals.

The importance of having a Digital Transformation strategy cannot be overstated. During an era of marked and quickened technological evolution and intense competition, businesses that fail to adapt and transform are at a significant risk of becoming irrelevant. A well-crafted strategy provides a roadmap for transformation, identifying critical areas for digital innovation and outlining the steps necessary to achieve it. It ensures that the organization's digital efforts are coherent, focused, and aligned with its long-term goals. Moreover, this strategy facilitates a culture of continuous improvement and innovation, enabling the company to take advantage of new opportunities and navigate challenges and disruptions as they arise.

Accomplishing a Digital Transformation strategy requires a multifaceted approach. It starts with a clear vision and commitment from the top leadership, recognizing that Digital Transformation is a strategic imperative that must be integrated into the company's foundation. This vision should then be distilled into actionable goals, objectives, and principles that are specific, measurable, achievable, relevant, and time-bound (SMART). Leadership must ensure top-down accountability, where the Digital Transformation vision is supported and prioritized at all levels of management, and bottom-up responsibility, encouraging innovation and digital initiatives within all areas of the company.

Central to executing a Digital Transformation strategy is establishing a Digital Transformation Office (DTO) or a similar entity that oversees the digital initiatives across the organization. This centralized coordination body is crucial in ensuring the Digital Transformation efforts are strategic, coherent, and aligned with the company's mission, objectives, and goals. The DTO fosters collaboration among various departments, leverages synergies between different digital initiatives, and ensures that the Digital Transformation efforts translate into tangible business outcomes.

Moreover, the strategy should emphasize building solid partnerships with technology providers, fostering a culture of innovation, enhancing customer experience through digital channels, and continuously improving operational efficiency through digital solutions. Engaging employees in this process through training and development in digital skills is also vital for fostering an adaptive and innovative organizational culture.

As businesses navigate the complex Digital Transformation journey, the outline for success lies in a carefully designed Digital Transformation strategy that deeply and strategically is integrated into every business facet, enabling organizations to pivot from antiquated and siloed practices to innovative, technology-driven processes and models.

This strategy is not just about adopting new technologies but about rethinking and radically changing how a business operates and delivers value to its customers. This strategic planning is critical because it aligns technological advancements with the company's business objectives, ensuring that every tech investment contributes directly to its mission, vision, and overarching goals.

Case Study: The Catalyst for Change

The boardroom was tense, the air thick with frustration. The senior executive suite of ConglomeraTech Inc., a global conglomerate operating across multiple industries, grappled with a persistent challenge that had eluded them for years.

"People, our operational costs continue to soar, and our efficiency metrics are stagnant at best," announced the CEO, his brow furrowed. "Despite our substantial investments in automation and Artificial Intelligence, we're not seeing the needle move as we had anticipated."

The room was silent, save for the occasional shuffling of papers and the tapping of pens on the polished mahogany table. Each executive knew the gravity of the situation – their competitors were outpacing them, threatening ConglomeraTech's market dominance.

The Chief Operating Officer spoke up, "We've been preaching the need for automation and process optimization for years, but something's not clicking. It's as if our efforts are scattered, failing to coalesce into a cohesive strategy." The Chief Information Officer nodded in agreement, "Our investments in AI and automation have been piecemeal, Point Solutions that address specific pain points but fail to deliver a holistic transformation."

The discussions continued, with each executive voicing their concerns and theories, but no clear path forward emerged. It was a recurring pattern, one that had plagued the company for far too long. Then, a voice from the back of the room cut through the din. It was Jennifer, a senior manager who had joined the company a year prior, bringing a wealth of Digital Transformation experience.

"Forgive my intrusion, but I believe I may have uncovered the root of our challenges," she began, her words carrying a weight that commanded attention.

During a recent "skip-level" meeting, where Jennifer had the opportunity to engage with employees outside her direct chain of command, she discovered a startling revelation. ConglomeraTech lacked a comprehensive Digital Transformation strategy that aligned with the company's mission, objectives, and goals. "Our Point Solution thinking has made us less competitive and more expensive to operate," Jennifer explained. "While our competitors have embraced a System Solution approach, taking a holistic view of their Digital Transformation, we've been stuck in a cycle of reactive, piecemeal investments."

A hush fell over the room as the executives absorbed Jennifer's words. It was a harsh truth but one that resonated deeply. The CEO leaned back in his chair, a glimmer of understanding in his eyes. "Jennifer, you may have just provided the catalyst we've been seeking," he said, his voice laced with a newfound determination.

From that moment on, ConglomeraTech embarked on a journey to become a truly Digitally Native firm. They assembled a cross-functional team, led by Jennifer, tasked with developing a comprehensive Digital Transformation strategy that would align their technology investments with their business objectives. It was an arduous process, requiring a deep dive into the company's operations, processes, and culture. However, with unwavering commitment and a clear vision, ConglomeraTech began to shed its inefficient, reactive approach to technology adoption.

Siloed systems gave way to integrated platforms, legacy processes were streamlined and automated, and a culture of continuous improvement took root. The company's workforce was upskilled and empowered to embrace the digital age enthusiastically and confidently. As the months passed, the impact was undeniable. Operational costs plummeted, efficiency soared, and ConglomeraTech regained its competitive edge. The executive suite, once mired in frustration, now beamed with pride, their belief in the power of Digital Transformation solidified.

At the heart of this transformation stood Jennifer, the catalyst who had sparked a revolution. She guided ConglomeraTech into a future where technology was not merely a tool but a strategic imperative woven into the fabric of the company's essence.

Introduction Key Points Recap: Digital Transformation and Intelligent Automation as Existential Needs of the Enterprise

- **The Evolution of RPA to Intelligent Automation:** RPA started as a means to automate rule-based, repetitive tasks, effectively mimicking human interactions with digital systems. Over time, it has evolved into Intelligent Automation, blending traditional RPA capabilities with AI, NLP, and ML. This transition marks a significant leap in operational optimization, efficiency enhancement, and service innovation.

- **Enterprise Automation and the Role of Governance Frameworks:** The introduction emphasizes the importance of adopting a holistic approach to automation, referred to as Enterprise Automation. This approach acknowledges the need for a cultural, organizational, and operational change across the entire enterprise, facilitated by the establishment of governance frameworks and Centers of Excellence (CoEs) to align IA initiatives with business strategies and goals.

- **IA During the COVID-19 Pandemic**: The pandemic highlighted the indispensable role of IA in enabling businesses to adapt to new challenges, such as the shift to remote work. Companies with pre-existing automation infrastructures, particularly those with an RPA Center of Excellence, were better positioned to wade through the disruptions caused by the pandemic, showcasing the value of foresight and strategic investment in automation technologies.

- **The Imperative of a Digital Transformation Strategy**: A coherent and comprehensive Digital Transformation strategy is paramount for successfully integrating technology into all aspects of a business. This strategy serves as a blueprint for radical changes in business operations, driving agility, efficiency, and a digital-first mindset. It underscores the need for top-down commitment and bottom-up engagement to drive digital initiatives.

- **The Story of ConglomeraTech:** This illustrative narrative highlights the transformative power of a strategic approach to digital adoption. By moving away from piecemeal investments towards a holistic Digital Transformation strategy, ConglomeraTech illustrates how businesses can regain their competitive edge, reduce operational costs, and embrace a culture of continuous innovation and improvement.

This introduction is just a brief summary of what's to come. It will give you a high-level understanding of the promise and challenge of prioritizing Digital Transformation for your firm and the central role Intelligent Automation can play. Now, let's dig a little deeper, and I promise you that the journey ahead will be informative and engaging.

Part 1: The Strategic Imperative of Intelligent Automation

Chapter 1: Intelligent Automation as a Driver of Business Model Innovation

Reshaping Business Models with Strategic IA Implementations

In the ever-evolving business landscape, organizations across industries are embracing Intelligent Automation as a catalyst for transformative change. By strategically implementing IA solutions, companies are reshaping their business models, unlocking new levels of efficiency, agility, and a focus on the customer. This picture explores how diverse organizations have leveraged the power of IA to redefine their operations, processes, and value propositions, ultimately driving their Digital Transformation journeys forward.

The world of professional services has long been characterized by labor-intensive, complex processes that often hinder efficiency and scalability. However, EY recognized the potential of IA to revolutionize its business model. Through the implementation of IA technologies, this major consulting firm streamlined its internal operations, automating tasks such as data entry, document processing, and report generation. This enabled their highly skilled consultants to focus on higher-value activities, such as customer and product analysis and client advisory services (UiPath, n.d.).

Moreover, this consulting firm leveraged IA to enhance its service delivery model. The firm could provide more tailored client recommendations by integrating AI-powered IA solutions into their client engagements. This shift towards a more technology-enabled, data-driven consulting model not only increased operational efficiency but also positioned the firm as a leader in the Digital Transformation space, better equipped to guide clients through their own transformation journeys.

In healthcare, IA has played a pivotal role in reshaping the patient experience and streamlining clinical processes. One notable example is Omega Healthcare, a major Asian Healthcare provider based in India, which has embraced IA to enhance patient care and operational efficiency (UiPath, n.d.). By deploying IA solutions, this renowned healthcare provider automated a range of administrative tasks, such as appointment scheduling, medical record management, and billing processes. This automation has reduced the administrative burden on healthcare professionals and improved the patient experience by minimizing wait times and ensuring accurate and timely information flow. This healthcare giant has also leveraged AI-powered diagnostic tools and predictive analytics to support clinical decision-making, leading to more personalized and effective patient treatment plans.

IA, too, has proven to be a game-changer for logistics and supply chain management. By integrating IA into its operations, the global shipping giant DHL has optimized routing and delivery schedules, resulting in substantial cost savings and enhanced customer service (DHL, n.d.). This shipping firm can use Advanced Analytics and Machine Learning to predict package volumes, identify potential bottlenecks, and adjust its operations dynamically using RPA to ensure efficient and timely deliveries. This data-driven approach has enhanced operational efficiency and enabled the company to offer customers real-time tracking and predictive delivery updates, strengthening its competitive advantage in the logistics industry.

Moving beyond these specific examples, IA has played a pivotal role in enabling the broader Enterprise Automation (EA) concept. EA represents a holistic approach to automation, where organizations integrate IA technologies across all levels and functions, aligning automation initiatives with overarching business strategies and goals sustainably and robustly.

One organization that has embraced the principles of EA is a telecommunications company based in West Asia, Vodafone Turkey. By establishing a centralized automation Center of Excellence (CoE), this telecommunication behemoth has been able to coordinate and govern its automation efforts across various business units (UiPath, n.d.). This centralized approach has enabled the company to identify synergies, leverage best practices, and ensure consistency in its IA implementations. The company also has leveraged IA to automate various processes, from customer service and billing to network operations and IT support. This has improved operational efficiency and enabled the company to deliver superior customer experiences by providing faster response times and proactive issue resolution. Furthermore, by aligning its IA initiatives with its broader Digital Transformation strategy, the telecommunications giant has been able to encourage a culture of continuous improvement and innovation. The company actively encourages employees to identify automation and process optimization opportunities, empowering them to drive positive change within the organization.

Unlocking Value through Intelligent Automation: A Catalyst for Transformation

As companies pursue competitive advantage and long-term sustainability, organizations across industries are turning to Intelligent Automation (IA) as a transformative force. By strategically harnessing the power of IA, companies are not only optimizing their operations and controlling expenses but also fostering innovation, revenue growth, and market differentiation. This narrative explores how diverse organizations are leveraging IA to create value across multiple dimensions, ultimately propelling their Digital Transformation journeys forward.

Proctor and Gamble, a leading consumer packaged goods (CPG) company, has harnessed IA to drive expense control and operational efficiency across its supply chain and manufacturing processes (P&G, n.d.). By deploying IA solutions, the company has automated processes such as those related to order processing, inventory management, and production scheduling. This has reduced manual workloads, minimized errors, and improved overall efficiency. By reallocating resources previously dedicated to these tasks, the CPG giant has been able to focus on more strategic initiatives, ultimately enhancing its bottom line.

Lufthansa, a major European airline carrier, has leveraged IA along with its extensive buildout of its cloud capabilities, to enhance its customer service offerings and drive revenue growth in the travel and hospitality industry (Technology Magazine, 2023). The airline has provided personalized assistance, recommendations, and tailored offers by integrating AI-powered chatbots, virtual assistants, and predictive analytics into its customer support channels. This has improved customer satisfaction and created new opportunities for upselling and cross-selling, generating additional revenue streams. Additionally, the airline has utilized IA to optimize flight scheduling, route planning, and aircraft maintenance, enabling it to deliver superior service quality and operational efficiency, further differentiating itself in a highly competitive market.

The prominent law firm, Barry, Appleman and Leiden (BAL) LLP, has embraced IA as a catalyst for rapid innovation and market differentiation within the legal services sector (UiPath, n.d.). By integrating AI-powered analytics and automation into its service delivery model, the firm has accelerated the development of new, technology-enabled solutions tailored to each client's unique legal needs. This agility and responsiveness have enabled the firm to iterate on its offerings and maintain a competitive edge. Furthermore, by streamlining internal processes such as legal research, document review, and case management through IA, the firm has improved operational efficiency, freeing up valuable time and resources for its attorneys to focus on higher-value activities like strategic legal advice and representation.

As these examples illustrate, strategic IA implementation has enabled organizations to unlock value across expense control, revenue growth, and market differentiation. IA's true power lies in its integration with an organization's overall Digital Transformation strategy – the holistic reimagining of processes, customer interactions, and corporate strategy through digital innovation.

IA serves as a critical enabler, providing tools to automate decision-making, optimize processes, and generate insights driving innovation and competitive advantage. Aligning IA with Digital Transformation goals ensures automation efforts contribute to a comprehensive approach to leveraging technologies. This alignment fosters continuous improvement and cross-functional collaboration and extends IA's benefits beyond operational efficiency to drive business growth, customer satisfaction, and sustainability.

As the digital age's complexities intensify, organizations embracing strategic IA within their transformation journeys will thrive. By unlocking value through expense control, revenue growth, and differentiation, they'll secure competitive edges and industry leadership.

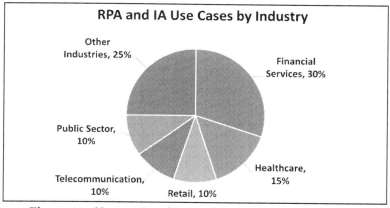

Figure 1.1 *(OpenAI, n.d.)*, **Distribution of RPA and IA use cases**

How Intelligent Automation Empowers Organizations to Thrive in a Rapidly Changing Landscape

In a world defined by constant change, competitive agility is the key to survival. Organizations must adapt quickly to shifting market conditions, customer demands, and emerging technologies to stay ahead of the curve. Intelligent Automation plays a pivotal role in fostering this agility, empowering organizations across industries to pivot strategically and seize new opportunities. Let's explore how IA fuels agility in scenarios that span several distinct industries and types of companies.

Consider a large non-profit organization focused on environmental conservation. This organization operates within a complex landscape of shifting public sentiment, fluctuating donor support, and ever-evolving environmental policies. Adapting to these dynamic factors is key to effectively driving their mission, but they're often hampered by manual processes that slow their ability to react. Enter the transformative power of IA. Tools powered by Intelligent Automation monitor news, social media, and relevant legislation for trends and changes impacting their field. These insights streamline their advocacy efforts, tailoring messaging based on emerging issues or changes in public perception. Within fundraising, ML-powered IA analyzes past donation patterns, identifying potential major donors and suggesting personalized outreach campaigns. These IA recommendations optimize fundraising efforts, ensuring resources are allocated effectively to maximize their impact. As a result, the non-profit becomes more agile, quickly pivoting its advocacy messaging to maximize its relevance, while also optimizing its fundraising strategies to ensure the resources necessary to drive its mission forward.

Shifting focus, let's look at a national hospitality chain with hotels and resorts across diverse locations. This company faces the challenge of catering to highly dynamic customer preferences, local events, and seasonal demand fluctuations. Legacy systems and reliance on manual analyses make adjusting their offerings for each location and time period time-consuming. IA becomes a core solution for infusing agility throughout their operations. IA-powered tools analyze vast datasets of travel trends, local happenings, weather patterns, and past customer data at each location. This analysis enables dynamic pricing strategies at both the room level and for on-site services like dining and activities. IA tools can even automate adjustments to their marketing messages based on local events or weather, ensuring they're attracting the right guests at the right time. In guest services, IA-powered chatbots handle routine requests like room upgrades or restaurant reservations, while sentiment analysis flags potential negative experiences for human staff to proactively address. The result?

A highly responsive hospitality chain that can maximize revenue through localized dynamic pricing, tailor marketing for increased occupancy, and provide exceptional guest experiences that drive loyalty. This agility leads to increased market share and customer satisfaction.

Now, let's switch to the field of higher education. Imagine a large university with diverse academic programs, a sprawling campus, and evolving student needs. Legacy systems and paperwork-heavy processes often create bottlenecks in student services, research administration, and even optimizing campus operations. IA offers a transformational solution. In student services, IA-powered chatbots can handle a wide array of common inquiries, from tuition deadlines to course registration, freeing up staff for complex issues. Within research administration, IA streamlines grant proposal processes, matching potential funding sources to faculty research and even pre-checking compliance with complex funder regulations. For facilities management, IA can analyze historical energy usage patterns and maintenance logs to optimize resource allocation and predict potential equipment failures, proactively preventing costly disruptions. The university, as a result, becomes far more agile. Students receive faster responses to their inquiries, researchers can focus their time on groundbreaking research, and the campus operates with greater efficiency. This focus on agility benefits the university's core mission across the board, ultimately leading to an enhanced academic experience and more efficient operations.

Across these probable examples, spanning diverse sectors, IA's transformative role in Digital Transformation shines brightly. Digital Transformation, at its core, is about embracing technology to reshape business fundamentally. IA weaves itself into the fabric of this transformation by automating tasks, extracting insights from vast datasets, and accelerating decision-making. This newfound ability to swiftly adapt to changing conditions, continually optimize processes, and provide exceptional experiences is what allows businesses not merely to survive but thrive in a world of continual digital disruption and competitive pressure.

Case Study: A Tale of Digital Transformation and Financial Performance

Max Return, a seasoned Wall Street investment analyst, had always believed in the transformative power of technology. As he sifted through the quarterly earnings reports of Card More and Distinguished Express, two financial services spin-offs from industrial conglomerates, the divergence in their strategies and outcomes couldn't have been starker. Max had closely monitored their journeys over the last eight quarters since their initial public offerings (IPOs), drawing insights that would soon shape the investment community's view of Digital Transformation in the financial sector.

Card More, under the leadership of Charles Chang, its Chief Digital Transformation Officer (CDTO), had embarked on an ambitious journey to become a Digitally Native company. This commitment was evident from his early decision to establish an Intelligent Automation (IA) Center of Excellence (CoE) and to coordinate emerging technology efforts under a cohesive Digital Transformation Strategy. This strategy aimed to reimagine the two companies' operations to be more efficient and effective, which was necessary to thrive as independent entities post-spin-off. Initially, this path was met with skepticism from the board of directors, who balked at the upfront costs associated with such a transformative agenda. However, Charles Chang had been persuasive, promising that the initial investment would yield exponential returns in efficiency, customer satisfaction, and, ultimately, revenue growth.

In contrast, Distinguished Express had taken a markedly different approach. Its leadership, eager to demonstrate quick wins to investors and the market, opted against a comprehensive Digital Transformation. While they did establish an RPA team, it lacked strategic alignment with broader business objectives, functioning more as a band-aid solution to operational inefficiencies than as a catalyst for genuine innovation.

As Max delved into the financials and operational reports, the results spoke volumes. Despite its initial high costs for setting up the IA CoE and other digital initiatives, Card More had begun to reap substantial benefits. By the third quarter post-IPO, the company reported a significant decrease in operational expenses, attributable to the efficiencies gained from automation and AI-driven insights. Revenue growth had outpaced projections, driven by an enhanced customer experience that attracted new clients and deepened engagement with existing ones. The exponential decrease in expenses and multiple increases in revenue validated Charles Chang's vision, transforming initial skepticism into widespread acclaim within the boardroom and beyond.

On the other side, Distinguished Express's financial health told a cautionary tale. Without a strategic foundation for its RPA initiatives, the company saw marginal short-term gains in operational efficiency but nothing that could contribute to sustainable growth or competitive advantage. Its costs, initially under control, began to creep up as patchwork solutions failed to address underlying inefficiencies. Revenue growth stalled, as the company could not offer the innovative services or seamless customer experience that today's market demands. By the sixth quarter, the stark reality set in with the lack of investment in true Digital Transformation hindering growth and now threatening the company's independence. The board began to explore desperate measures, including a potential sale to or merger with a competitor.

The irony of Distinguished Express's potential acquirer being Card More was not lost on Max. It epitomized the ultimate vindication of Card More's digital-first strategy and served as a stark warning to companies hesitant to embrace Digital Transformation. As Max prepared his analysis for the next investment newsletter, this story centered on the pivotal role of strategic Digital Transformation in defining the future of financial services. Through the lens of Card More's success and Distinguished Express's challenges, he highlighted how a coherent, forward-looking digital strategy, exemplified by an IA CoE aligned with corporate vision, could safeguard a company's independence and propel it to industry leadership.

Max's commentary extended beyond mere financial metrics; it delved into the strategic imperatives that would dictate the survivability and competitiveness of financial services firms in a digital age. His insights illuminated the broader implications for the industry, emphasizing the necessity of being Digitally Native to navigate the complexities of the modern financial landscape successfully. As executives and investors digested Max's analysis, the story of Card More and Distinguished

Express became a guiding narrative for the future of Digital Transformation in the financial sector.

Chapter 1 Key Points Recap: Intelligent Automation as a Driver of Business Model Innovation

- **Strategic IA Implementations in Professional Services:** A global consulting firm capitalized on IA to streamline internal processes and enhance service delivery, freeing consultants to focus on high-value activities and offering data-driven insights to clients. This transformation reinforced the firm's leadership in guiding Digital Transformation.
- **IA's Role in Healthcare Transformation:** A leading healthcare institution utilized IA to automate administrative tasks and integrate AI-powered diagnostic tools, significantly improving patient care and operational efficiency. This showcases the potential of IA in enhancing the patient experience and clinical decision-making.
- **Optimizing Logistics with IA:** A global shipping giant implemented IA to optimize routing and delivery schedules, demonstrating substantial cost savings and improved customer service. Advanced Analytics and

Machine Learning algorithms enabled dynamic operations adjustments and real-time customer updates, highlighting IA's impact on operational efficiency and customer satisfaction.

- **Enterprise Automation (EA) as a Holistic Approach:** The chapter illustrates EA's significance through the case of a European telecommunications company that established a Center of Excellence (CoE). This approach enabled the identification of synergies, leveraging best practices, and ensuring consistency in IA implementations, driving superior customer experiences and fostering a culture of continuous improvement and innovation.

- **Expanding IA's Impact Beyond Operational Efficiency:** This demonstrates and underscores IA's ability to drive expense control, revenue growth, and market differentiation when aligned with an organization's Digital Transformation strategy through examples from various sectors, including consumer packaged goods, travel and hospitality, legal services, and retail.

- **Fostering Competitive Agility with IA:** This showcases IA's critical role in enabling organizations to quickly adapt to market changes, using examples from the non-profit, hospitality, and education sectors. IA fuels agility by automating tasks, generating actionable insights, and accelerating decision-making, allowing organizations to pivot strategically and seize new opportunities.

- **Digital Transformation and Financial Performance – Max Return's Observations:** The contrasting strategies and outcomes of two financial services spin-offs, Card More and Distinguished Express, underscore the pivotal role of a coherent, strategic approach to Digital Transformation. Card More's success story, rooted in its commitment to becoming Digitally Native and establishing an IA CoE, emphasizes the importance of aligning IA initiatives with the corporate vision to achieve sustainable growth, efficiency, and competitive advantage.

This chapter illustrates the transformative potential of IA in redefining business models and enhancing competitiveness across industries. It emphasizes the necessity of a strategic, holistic approach to Digital Transformation, aligning IA implementations with broader business objectives.

In Chapter 2, we will discuss how Intelligent Automation can directly and positively impact your customers.

Chapter 2: Transforming the Customer Experience with IA

Using IA in Delivering Proactive Customer Service on a Massive Scale

Intelligent Automation enables more complex, decision-based processes by integrating advanced technologies such as Machine Learning, Natural Language Processing, and Artificial Intelligence. This shift towards a more Intelligent Automation framework is pivotal for companies aiming to enhance customer experiences, streamline operations, and foster a Digitally Native culture.

At the core of IA's application in customer service is its ability to deliver proactive, personalized interactions at scale. By analyzing vast amounts of data in real-time, IA solutions can predict customer inquiries or issues before they arise, offering solutions or information preemptively. This capability transforms the customer service paradigm from reactive to proactive, significantly enhancing customer satisfaction and loyalty. Based on the data it was trained with, for instance, an ML-driven IA solution can anticipate potential issues with a product or service, initiating contact with the customer to address these issues before they escalate into more significant problems.

Moreover, IA's integration into customer service operations, mainly through attended automation, revolutionizes the interaction between customer service representatives (CSRs) and customers. Attended Automation refers to IA solutions that work alongside human agents, providing them with real-time information, guidance, and support during customer interactions. This symbiotic relationship between humans and AI enables CSRs to deliver more accurate, empathetic, and efficient service. For example, during a call, an IA solution can instantly provide a CSR with a customer's full profile, purchase history, and previous issues, allowing the CSR to tailor the conversation and solutions offered to the customer's specific context and needs.

In contact centers, IA dramatically enhances the efficiency and effectiveness of customer service operations. By automating routine inquiries and tasks, IA allows human customer service representatives to emphasize more complex and esoteric customer issues. Furthermore, IA-powered chatbots and virtual assistants can simultaneously handle a large volume of customer questions, reducing wait times and improving overall service levels (QuantHub, n.d.). These intelligent solutions can resolve common issues directly or escalate more complex cases to human agents, ensuring that customers receive the appropriate focus and level of assistance (Chatbot, n.d.).

The application of IA extends beyond customer service interactions to include the automation of transactions and processes that directly impact the customer experience. In sectors such as banking and insurance, IA is revolutionizing how transactions like loan approvals and insurance claims are processed. Traditionally, these processes have been time-consuming, involving manual data entry, verification, and decision-making. IA accelerates these transactions by automating data collection, analysis, and processing, significantly reducing the time from application to approval or claim settlement.

For example, IA solutions can quickly assess a loan application by pulling data from various sources, evaluating creditworthiness, and deciding based on predefined criteria. This speeds up the loan approval process and ensures consistency and fairness in decision-making. Similarly, in insurance claims processing, IA can automate damage assessment, fraud detection, and payment calculations, expediting the settlement process and improving customer satisfaction during what is often a stressful time. In transactional processes, such as loan approvals or insurance claims, IA dramatically speeds up and enhances these experiences for customers.

Traditionally, such processes have been bogged down by manual interventions, lengthy paperwork, and slow decision-making. IA, however, introduces a paradigm shift by automating the entire process, from application processing to decision-making and final execution. In the context of approving and funding loans, for example, IA solutions can quickly assess a customer's creditworthiness, verify documents, and calculate risk factors using advanced algorithms. This reduces the time required to process loan applications, increases accuracy, reduces the risk of errors, and ensures a faster disbursement of funds to customers.

The adoption of IA in delivering proactive customer service and accelerating transactions is a critical component of a broader Digital Transformation strategy. By implementing IA, businesses improve efficiency, reduce costs, and create more fulfilling customer experiences. This alignment of technological innovation with strategic business goals and customer needs is at the heart of Digital Transformation.

Moreover, the application of IA in customer interactions and transactional processes generates vast amounts of data. This data, when analyzed, provides valuable insights into customer behavior, preferences, and trends, enabling businesses to tailor their products, services, and interactions more effectively to meet customer needs. This data-driven approach enhances the customer experience and gives businesses a competitive edge, as they can anticipate market changes and adapt more swiftly to evolving customer expectations.

Additionally, the journey towards becoming a Digitally Native company involves embedding digital technologies into all business areas, enabling rapid adaptation to changes in the technological landscape and customer behavior. The use of IA in customer service and transaction processing is a prime example of how companies can leverage digital technologies to innovate, compete more effectively in the digital age, and sustain long-term growth.

The Link between IA-Powered Customer Experience Initiatives and Increased Brand Loyalty

Moving beyond familiar territories like attended automation in contact centers, IA finds its essence in a myriad of applications that significantly enhance brand loyalty, advocacy, and, consequentially, revenue growth. This account delves into the profound impact of IA across various business operations, providing a nuanced exploration of its role in marketing, operations, customer service beyond traditional centers, and accelerating customer transactions, all while ensuring a seamless Digital Transformation journey.

One compelling example of IA's application lies in customer onboarding processes across industries. Consider the financial sector, where onboarding new clients is often fraught with paperwork and compliance checks. IA streamlines this process by automating document verification and data entry tasks, significantly reducing the time from application to account activation. This swift, efficient onboarding experience enhances customer satisfaction and builds trust and loyalty from the outset, encouraging customers to explore and adopt additional services.

Turning to the operational efficiencies within the supply chain, IA demonstrates its prowess in the retail and manufacturing sectors. By employing IA for demand forecasting and inventory management, businesses can dynamically anticipate market trends, adjust stock levels, and optimize logistics for delivery. An e-commerce company, for instance, could leverage IA to automatically adjust inventory based on real-time sales data and supplier lead times, ensuring product availability aligns with customer demand. This proactive approach minimizes stockouts and excess inventory, directly contributing to a positive customer experience and fostering brand loyalty.

Beyond operational processes, IA redefines customer service interactions through sophisticated, context-aware chatbots and virtual assistants. Unlike traditional chatbots that offer limited responses, IA-powered assistants can conduct nuanced conversations, providing customized recommendations and support. For example, a travel agency could deploy an IA solution that offers customized travel packages by analyzing the customer's preferences and previous interactions. This personalized engagement enhances the customer's experience and promotes higher conversion rates and customer loyalty.

Moreover, IA's influence extends to improving the efficiency and personalization of customer transactions. In sectors such as retail and hospitality, IA can tailor promotions and loyalty rewards in real-time based on the customer's purchasing history and preferences. This level of personalization, powered by IA's analytical capabilities and Machine Learning, not only elevates the shopping experience but also encourages repeated business and word-of-mouth promotion, amplifying brand loyalty and advocacy.

Case Study: The Story of First Metropolitan Bank

Valerie Blue had been a loyal customer of First Metropolitan Bank for over two decades, a relationship that had weathered the ups and downs of life's financial journey. However, in recent years, she found herself increasingly frustrated by the deteriorating level of customer service she had once taken for granted.

Lengthy phone wait times, miscommunications regarding her accounts, and a general sense of inefficiency had become the norm. A busy professional, Valerie found herself contemplating the unthinkable: leaving the bank she had trusted for so long in search of a more seamless and satisfactory experience elsewhere.

Yet, just as her patience wore thin, Valerie noticed a remarkable transformation unfolding. Her interactions with First Metropolitan Bank became smoother, more efficient, and genuinely customer-focused. From routine inquiries to more complex transactions, like her recent home line of credit, the bank's responsiveness and attention to detail were nothing short of impressive.

Intrigued by this sudden improvement, Valerie tuned in to a segment on the International Business News (IBN) channel, where First Metropolitan Bank's CEO discussed the company's Digital Transformation efforts. She listened intently as the CEO highlighted the bank's strategic investments in Intelligent Automation technologies as a critical component of its customer service and operational overhaul.

"At First Metropolitan, we recognized the need to adapt to the changing digital landscape and meet our customers' evolving expectations," the CEO explained. "By leveraging the power of Intelligent Automation, we have streamlined our processes, enhanced operational efficiency, and empowered our employees to deliver exceptional customer experiences."

As Valerie listened, the pieces began to fall into place. The seamless funding of her home line of credit, the prompt and knowledgeable responses from customer service representatives, and the overall sense of efficiency she had recently experienced resulted from First Metropolitan Bank's commitment to Digital Transformation and the integration of Intelligent Automation into their operations.

Valerie recalled the frustration she had felt just months earlier, contemplating the daunting task of switching banks after decades of loyalty. Now, she found herself appreciating the bank's proactive efforts to address the very issues that had once driven her to consider such a drastic step.

The CEO's words resonated deeply: "We understand that in today's fast-paced world, our customers demand reliable financial services and a seamless, unique personal experience. By harnessing the power of Intelligent Automation, we are improving our operational efficiency and ability to truly understand and anticipate our customers' needs."

As the segment concluded, Valerie felt renewed confidence in her longstanding relationship with First Metropolitan Bank. The bank's willingness to embrace Digital Transformation and leverage cutting-edge technologies like Intelligent Automation demonstrated its commitment to staying ahead of the curve and delivering exceptional customer service.

At that moment, Valerie knew that her loyalty had been well-placed. She looked forward to a future where her banking experiences would be defined by efficiency, personalization, and the bank's unwavering dedication to putting its customers first.

Chapter 2 Key Points Recap: Transforming Customer Experience with IA

- **Proactive Customer Service at Scale:** IA's real-time data analysis enables a shift from reactive to proactive customer service, significantly boosting customer satisfaction and loyalty by preemptively addressing potential issues.
- **Enhanced Customer Service Representative (CSR) Interactions:** Through attended automation, IA aids CSRs with immediate data and guidance, fostering more personalized, efficient, and empathetic customer interactions.
- **Increased Efficiency in Contact Centers**: Automating routine inquiries and tasks, IA frees up human agents for complex customer issues, enhancing

contact center operations' overall efficiency and effectiveness.

- **Streamlining of Transactional Processes:** IA expedites critical transactions like loan approvals and insurance claims by automating data collection, analysis, and processing, leading to quicker and more accurate decisions.

- **Data-Driven Insights for Tailored Customer Experiences:** IA generates valuable data from customer interactions and processes, offering insights that help tailor products and services to effectively meet customer needs.

- **Foundation for Digital Transformation:** Implementing IA to enhance customer service and streamline transactions aligns with broader Digital Transformation goals, emphasizing agile, customer-focused operations that leverage digital technologies.

- **Story of Valerie Blue:** The narrative of Valerie Blue and First Metropolitan Bank exemplifies IA's transformative impact. Valerie's renewed faith in the bank, following its Digital Transformation efforts that significantly improved her customer experience, illustrates the tangible benefits of integrating IA in business operations. This story underlines the importance of Digital Transformation and IA in retaining customer loyalty and adapting to the digital age.

This chapter explains IA's crucial role in redefining customer service and transactional efficiency, setting a foundation for strategic Digital Transformation. It showcases how governance, foresight, and a unified strategy are essential to success.

In Chapter 3, we will examine how IA is empowering employees throughout organizations with enhanced abilities - increasing their productivity and their value to their firms.

Chapter 3: IA as an Employee Empowerment Tool

Addressing and Proactively Countering Concerns About IA-Induced Job Displacement

Integrating Intelligent Automation (IA) into the modern workplace is a transformative movement reshaping how businesses operate across various industries. While the efficiency and productivity gains from IA are undeniable, its rise brings concerns about job displacement to the fore. This topic resonates deeply within the fabric of organizational culture and worker sentiment. As we delve into this conversation, it's essential to understand that the journey of IA, particularly within the confines of business office processes, is not a tale of replacement but one of empowerment, evolution, and strategic opportunity for both businesses and their employees.

In the realm of process automation, spanning Finance, HR, Customer Service, and beyond, IA has carved its niche not as a usurper of roles but as a catalyst for Digital Transformation. This transformation is predicated on the notion that automating routine, repetitive tasks frees human capital to engage in higher-value activities that demand creativity, critical thinking, and emotional intelligence, qualities, at least at its current stage of development, that IA cannot replicate.

Consider a financial services company grappling with the high volume of routine transactions and data processing tasks. Initially, the introduction of IA to handle processes such as document verification, data entry, and compliance checks sparked fears of job loss among the workforce. However, as the IA program matured, it became evident that the automation of these processes did not result in job displacement. Instead, employees were retrained and redeployed into roles focused on customer engagement, financial analysis, and strategy development, areas where human insight and interaction were paramount.

This evolution underscores a vital facet of Digital Transformation: the shift towards a more strategic, innovative workforce that leverages IA to enhance service delivery and create new value propositions. For instance, in the case of the financial services company mentioned before, customer service representatives transitioned from mundane query handling to roles that emphasized personalized financial advice, leveraging insights generated by IA tools to tailor solutions to individual customer needs.

Moreover, the story of IA and job displacement is also a story of organizational learning and adaptability. As businesses integrate IA into their operations, they embark on a continuous learning journey, identifying new opportunities for process improvement and workforce development. For example, a healthcare provider that implemented IA to manage patient records and appointment scheduling exemplified this learning journey. Automating these processes led to the realization that the healthcare provider's staff could be better utilized in patient care and support roles, thereby enhancing the patient experience and improving care outcomes.

The proactive counter to concerns about IA-induced job displacement lies in the strategic approach to workforce planning and development. Businesses that succeed in this endeavor view their employees as partners in Digital Transformation, investing in their upskilling and reskilling to navigate the new landscape shaped by IA. This approach mitigates the fears associated with job displacement and fosters a culture of innovation and continuous improvement. This centralized collaboration plays a crucial role in managing change, facilitating cross-functional collaboration, and driving the cultural shift necessary to embrace IA as an enabler of growth and innovation.

It's clear that IA's role extends beyond the automation of tasks to reshaping business models, enhancing client experiences, and opportunities for growth. By focusing on the human aspect of automation, businesses can navigate the challenges associated with IA-induced job displacement, turning potential disruption into a strategic advantage. Through a concerted effort that encompasses leadership commitment, workforce engagement, and a clear vision for the future, organizations can harness the full potential of IA to propel their Digital Transformation journey forward, ensuring that they not only survive but thrive in the ever-evolving business landscape.

Highlighting how IA Elevates the Workforce

Traditionally, employees across sectors have been bogged down by repetitive tasks such as data entry, invoice processing, and routine customer inquiries. These tasks, while necessary, consume significant amounts of time and energy, detracting from more strategic, creative work that can add real value to a company. IA steps in as a powerful solution, automating these routine tasks with unparalleled speed and accuracy. By taking over such duties, IA acts as a liberating force, freeing employees to concentrate on complex problem-solving, decision-making, and innovation. This shift not only enhances job satisfaction and engagement but also propels companies forward by harnessing the full potential of their human capital.

Further exploring IA's role, we delve into its dual functionality as both a digital assistant and teammate. In scenarios where human interaction is essential, attended automation comes into play. Here, IA operates alongside employees, acting as a digital assistant that supports real-time decision-making. For instance, attended bots can provide agents instant access to customer data, suggest personalized service options, or automate the completion of forms during a live customer call. This symbiosis between humans and machines enhances service delivery, reduces errors, and speeds up response times, all while ensuring that the human touch remains integral where it is most valued (Daugherty, 2018).

On the other side of the spectrum, unattended automation showcases IA's capability to function as a digital teammate, autonomously executing tasks without the need for human intervention. This is particularly transformative in back-office operations, such as the automatic reconciliation of accounts payable and receivable. Here, IA works behind the scenes, processing invoices, matching payments, and updating ledgers around the clock. This accelerates the financial close process and significantly reduces the potential for human error, showcasing IA's role as a dependable and efficient teammate that ensures business continuity and operational excellence.

As an assistant and teammate, the deployment of IA significantly furthers Digital Transformation initiatives. By automating processes, organizations can achieve unprecedented levels of efficiency and accuracy, paving the way for a more agile and responsive business model. This transformation is not just about streamlining operations; it's about reimagining how work is done. As routine tasks are automated, organizations are compelled to rethink their business processes, organizational structures, and even the nature of the work itself. This introspection often leads to more profound changes, developing a continual improvement and innovation culture.

Emphasizing the Role of IA in Employee Upskilling Initiatives and Long-Term Employee Development

The introduction of IA into the workplace offers employees an invaluable opportunity to upskill, shifting their focus from routine, manual, repetitive work to more strategic, value-added activities, which are often more meaningful from a human perspective. This shift is not merely about technical proficiency but encompasses a broader understanding of how technology can drive business innovation and efficiency while meaningfully changing the company's culture.

The role of IA in employee upskilling initiatives is multifaceted, involving both the direct upskilling of employees to deploy, manage, and optimize IA solutions and the indirect upskilling that occurs as employees are freed from routine tasks. For instance, consider a financial services company where IA is deployed for customer due diligence, claims processing, and financial reporting processes. Employees, who traditionally spent considerable time on these tasks, can now focus on analyzing trends, identifying areas for process improvement, and developing strategies for enhancing customer satisfaction and operational efficiency. This not only elevates the role of employees but also instills a culture of innovation and continuous improvement.

The focus for employees in a Digitally Native firm extends beyond certain tools or technologies. It encompasses developing a digital-first mindset, characterized by agility, curiosity, and the ability to leverage digital solutions to solve complex business challenges. Employees should be encouraged to think about the Digital Transformation of their roles and responsibilities, understanding how IA can be applied to enhance their work and contribute to the firm's strategic objectives. This requires a commitment to lifelong learning, where employees are provided with continuous training and development opportunities to keep pace with evolving technologies and business models (Marco Iansiti K. R., 2020).

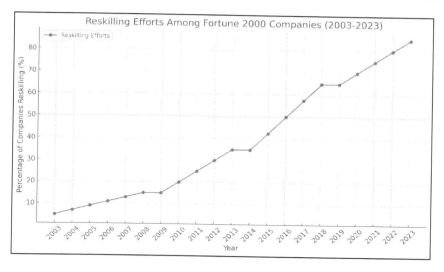

Figure 3.1 (OpenAI, n.d.), This graph depicts the reskilling efforts among Fortune 2000 companies from 2003 to 2023. It illustrates a significant increase in the percentage of these companies focusing on reskilling their employees, particularly as Digital Transformation efforts have expanded. This trend underscores the importance of workforce development in keeping pace with technological advancements and ensuring that employees are equipped with necessary new skills.

To support this, Digitally Native firms must invest in comprehensive training programs that cover technical skills related to IA and soft skills such as critical thinking, problem-solving, and adaptability. Such programs can be augmented with mentorship opportunities, cross-functional projects, and innovation labs, where employees can experiment with new technologies and develop solutions to real-world business challenges. These initiatives enhance employees' technical competencies and foster a culture of innovation, collaboration, and digital literacy across the organization (Saldanha, 2019).

In parallel, leadership plays a critical role in modeling the digital-first mindset, demonstrating a commitment to leveraging digital technologies to drive business success. Leaders should actively promote a culture of experimentation and learning, encouraging employees to explore new digital tools and approaches and recognizing and rewarding innovative ideas and initiatives. This sets the tone for a corporate culture that values digital innovation and continuous learning, ensuring the firm remains at the forefront of Digital Transformation.

The impact of IA on employee upskilling and long-term development is a critical component of a firm's Digital Transformation journey. By fostering a digitally literate workforce, organizations enhance their operational efficiency and competitive advantage and build a strong foundation for sustained growth and innovation. Employees, equipped with a deep understanding of digital technologies and their application in business, become agents of change, driving the firm's evolution in an increasingly digital world.

The role of IA in employee upskilling and development is integral to the Digital Transformation of firms, particularly those aspiring to become or maintain a Digitally Native status. By embracing IA, firms can cultivate a technically proficient, strategically focused, innovative, and adaptable workforce. This furthers the firm's Digital Transformation efforts and ensures its resilience and competitiveness in a rapidly evolving digital landscape. Through a commitment to continuous learning and development, supported by a culture that values innovation and digital fluency, Digitally Native firms can ensure their sustained success.

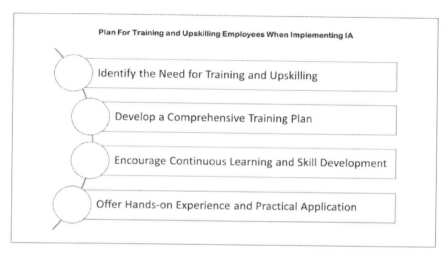

Figure 3.2 (FasterCapital, n.d.), Thing to consider for Upskilling Enterprise Talent

Case Study: The Story of Clair Riskless and Cher Lowbalance

Working diligently at a mid-size financial services firm, two analysts, Clair Riskless and Cher Lowbalance, were engrossed in a vital yet daunting task. Trusted with the responsibility of ensuring the accuracy of customer statements from their credit card partner, PayLater, their job was crucial for maintaining the firm's financial integrity. Their mission was to reconcile these statements against the balances reflected in the firm's payment platform, and a task easier said than done.

For Clair and Cher, each day, they were presented with a mountain of statements, each needing meticulous review. Before they could even begin the analytical work that drove value, they faced a Herculean task of downloading, organizing, and categorizing countless statements by region, retailer, and balance. This preliminary phase consumed a staggering 80% of their time, relegating the crucial analysis to the remaining 20%. Despite this limited capacity for in-depth review, their sample testing revealed a troubling trend: the firm was making overpayments to PayLater to the tune of $5 million a month. This discovery alone raised alarms, highlighting inefficiencies and financial leakages that demanded immediate attention.

The revelation led to a pivotal question: What could Clair and Cher uncover if the ratio of clerical work to analysis was reversed? With more time dedicated to analysis, they hypothesized the potential to recoup upwards of $25 million a month (or $300 million per year), a figure that could significantly impact the firm's financial health. The potential for such recovery underscored the critical need for change in their process.

Enter the solution that would revolutionize their roles and redefine the scope of their contributions: the deployment of an unattended automation teammate. This digital ally was designed to shoulder the burden of downloading and organizing statements, a task that had previously monopolized most of their time. With this automation in place, the landscape of their daily work transformed dramatically.

Freed from the chains of clerical duties, Clair and Cher could now immerse themselves in the core of their work, which was a deep, analytical review. The shift was monumental. Their productivity skyrocketed, and the value they brought to the firm saw a remarkable increase. They were no longer just analysts; they became strategic assets, uncovering insights and identifying financial discrepancies that had previously gone unnoticed.

This newfound efficiency allowed them to dive deeper into the data, uncovering subtle trends and anomalies that would have been impossible to detect with their previous time constraints. Their work became proactive rather than reactive, enabling the firm to safeguard against overpayments and optimize its financial processes.

For the firm's executives, the story of Clair and Cher became a powerful testament to the transformative potential of Intelligent Automation-led Digital Transformation. It showcased how strategic automation could unlock the true potential of their workforce, turning a routine task into an opportunity for significant financial savings and operational excellence. The story of their journey from being bogged down by manual tasks to becoming pivotal figures in financial analysis and strategic decision-making serves as an inspiring example for any organization looking to harness the power of Digital Transformation.

Chapter 3 Key Points Recap: IA as an Employee Empowerment Tool

- **Reframing the IA Narrative**: Rather than viewing IA as a threat to job security, we illustrate its role as a catalyst for employee empowerment and business innovation. Through automation, employees transition from mundane tasks to roles that demand and value their unique human capabilities.

- **Enhanced Workforce Strategy with IA:** By integrating IA, businesses transform their workforce strategies, promoting a shift towards roles focused on customer engagement, financial analysis, and strategic initiatives. Employees are retrained and redeployed in areas where human interaction and insight are crucial.

- **IA's Dual Functionality**: IA serves as both a digital assistant and a teammate. Attended automation supports real-time decision-making alongside employees, while unattended automation handles tasks independently, optimizing back-office operations and enhancing service delivery without human intervention.

- **Upskilling and Employee Development**: The introduction of IA presents opportunities for employee upskilling, shifting focus to strategic and creative work. It fosters a digital-first mindset among employees, emphasizing the importance of agility, curiosity, and leveraging technology for business solutions.
- **Leadership and Cultural Shift:** Integrating IA requires a commitment to a digital-first culture, continuous learning, and innovation. Organizations can effectively navigate the Digital Transformation journey by investing in employee development and fostering a culture that values digital literacy.
- **Transformative Impact on Financial Surveillance:** The story of Clair Riskless and Cher Lowbalance exemplifies IA's transformative impact. By automating clerical tasks, they shifted their focus to in-depth analysis, uncovering significant financial discrepancies and adding unparalleled value to their firm. Their journey highlights IA's potential to redefine roles and enhance operational efficiency.

This chapter emphasizes that the true power of IA lies in its ability to augment human capabilities, fostering a more innovative, strategic, and engaged workforce. By embracing IA, businesses can transform challenges into opportunities for growth, ensuring their resilience and success in a rapidly evolving digital landscape.

Moving forward in Part 2, we will be discussing how IA can be used to provide a strong foundation for Digital Transformation and will specifically discuss in Chapter 4 how to lay a roadmap to do so.

Part 2: Building a Foundation for Digital Transformation Success based on IA

Chapter 4: Developing a Strategic Roadmap for Intelligent Automation

The Importance of Aligning IA Initiatives with Overarching Digital Transformation Strategy

Within the modern enterprise, the synergy between Intelligent Automation (IA) and Digital Transformation strategies emerges as a pivotal axis around which the contemporary organization orbits. This explores the critical alignment of IA initiatives with an organization's overarching Digital Transformation strategy, shedding light on the tactical essence of IA against the strategic backdrop of Digital Transformation. Furthermore, it dissects the contrast between the Point Solution nature of IA and the System Solution approach inherent in Digital Transformation, articulating their definitions and roles within the larger scheme of enterprise evolution (Ajay Agrawal J. G., 2022). The discussion aims to unravel how these components interlace to propel the Digital Transformation journey, focusing on how Intelligent Automation is a chief enabling force.

At the center of Digital Transformation lies the strategic reimagination of business in the digital age. It's a broad, holistic endeavor that encompasses revamping business models, processes, and organizational structures to leverage digital technologies, thereby driving fundamental changes to how businesses operate and deliver value to customers. Digital Transformation is inherently strategic because it requires a long-term vision that aligns with the core objectives of an organization, necessitating leadership buy-in, cultural shifts, and significant investments in technology and skills development. Its scope transcends the boundaries of specific processes or functions, aiming instead to foster a Digitally Native enterprise that can adapt to and capitalize on the evolving digital landscape (Saldanha, 2019).

Conversely, as a tactical tool within this intricate vision, Intelligent Automation provides the means to operationalize the Digital Transformation strategy on the ground. However, the nature of IA as a Point Solution implies that its application is targeted and specific. It addresses distinct processes or activities within an organization, like automating data entry for invoice processing.

The dichotomy between the Point Solution approach of IA and the System Solution framework of Digital Transformation is stark yet complementary. Like a piece of a jigsaw puzzle, a Point Solution provides immediate, tangible benefits to a particular function or process. It's relatively quicker and less resource-intensive to implement, offering organizations a way to achieve quick wins and demonstrate the value of automation technologies. These initiatives, while beneficial, are narrowly focused and, without a broader strategy, risk creating silos that can impede overall business agility and efficiency, let alone transformation effectiveness.

On the other hand, the System Solution approach of Digital Transformation is akin to assembling the entire jigsaw puzzle (Ajay Agrawal J. G., 2022). It requires a comprehensive understanding of the current and future state of the business, meticulous planning, and coordination across various departments, functions, and lines of business. This approach seeks to integrate digital technologies throughout the organization, creating a seamless ecosystem that enhances operational efficiency, fosters innovation, and delivers superior customer experiences. By adopting a System Solution mindset, companies embark on a path of continual improvement and adaptation, leveraging digital capabilities to solve existing challenges and anticipate and prepare for future opportunities and threats (i.e., become Digitally Native).

Therefore, the alignment of IA initiatives with an enterprise's Digital Transformation strategy is crucial. It ensures that tactical automation projects are not isolated endeavors but are integrated into the broader vision of transforming the enterprise into a nimble, digitally savvy entity. This alignment facilitates the scaling of automations across the organization, enhancing their impact and ensuring they contribute to strategic objectives such as improving customer satisfaction, entering new markets, or developing innovative products and services.

Consider a financial services firm embarking on a Digital Transformation journey. The strategic goal might be to enhance customer experience and operational efficiency. In this context, IA can play a tactical role by automating manual, time-consuming processes like loan application processing or customer verification. However, for these initiatives to contribute meaningfully to the firm's Digital Transformation, they must be integrated with other digital initiatives, such as being a part of an initiative to revamp and streamline an entire business function digitally or leveraging RPA, Data Analytics, and Machine Learning in order to provide more customized offerings for its products. This requires a coordinated approach where IA projects are prioritized based on their strategic relevance and their ability to interlock with other digital technologies to create a cohesive, digitally transformed enterprise.

The interplay between the tactical deployment of Intelligent Automation and the strategic imperatives of Digital Transformation embodies the dynamic journey of modern businesses toward being Digitally Native. By harmonizing IA initiatives with the broader Digital Transformation strategy, organizations can ensure that their investments in automation technologies are not merely isolated projects but pivotal elements of a comprehensive effort to reimagine and revitalize their operations, culture, and customer engagement. This narrative underscores the essence of adopting a System Solution perspective, leveraging the targeted benefits of Point Solutions like IA to fuel the strategic ambitions of Digital Transformation.

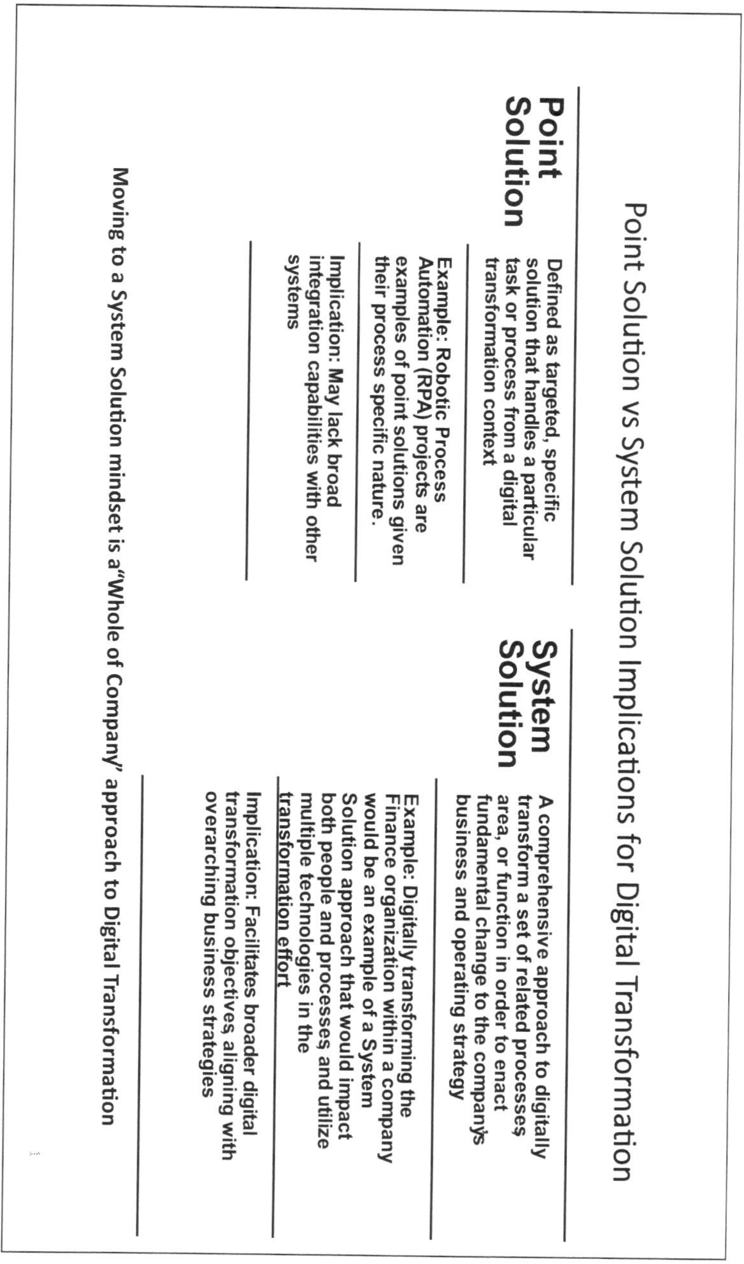

Point Solution vs System Solution Implications for Digital Transformation

Point Solution

Defined as targeted, specific solution that handles a particular task or process from a digital transformation context

Example: Robotic Process Automation (RPA) projects are examples of point solutions given their process specific nature.

Implication: May lack broad integration capabilities with other systems

System Solution

A comprehensive approach to digitally transform a set of related processes, area, or function in order to enact fundamental change to the company's business and operating strategy

Example: Digitally transforming the Finance organization within a company would be an example of a System Solution approach that would impact both people and processes, and utilize multiple technologies in the transformation effort

Implication: Facilitates broader digital transformation objectives, aligning with overarching business strategies

Moving to a System Solution mindset is a "Whole of Company" approach to Digital Transformation

Figure 4.1 (Ajay Agrawal J. G., 2022), Defining the difference between a Point Solution and System Solution

Developing a Practical Prioritization Framework for Identifying High-Potential Automation Opportunities

Developing a practical prioritization framework entails a multifaceted approach, integrating quantitative and qualitative assessments to discern the automation readiness and potential impact of various processes. This includes evaluating the complexity of the process, the stability of the underlying systems, and the expected return on investment. A crucial aspect of this framework is the alignment with the organization's strategic objectives, ensuring that automation initiatives propel the business toward its overarching goals of growth, customer satisfaction, and innovation. By adopting a methodical approach to prioritization, organizations can ensure a disciplined allocation of resources to initiatives that promise the highest payoff, thereby accelerating their Digital Transformation journey.

Integral to the successful implementation of IA is the development of an effective communication plan that addresses the multifaceted dimensions of IA within the context of Digital Transformation. Such a plan is pivotal in clarifying IA for the broader enterprise, delineating its capabilities, and setting realistic expectations. A cornerstone of this communication strategy is clearly articulating what IA is and, equally importantly, what it is not. Unlike traditional application development that typically involves creating bespoke solutions or systems from scratch, IA focuses on automating existing processes using RPA and possibly Machine Learning and other related technologies, to mimic human actions across the applications utilized by the processes currently being completed by people.

Effective communication underscores the distinction between automating a process and developing an application, as this is a critical differentiator that the organization must grasp in order for IA to be effective. This clarification helps mitigate misconceptions and confusion among stakeholders and the company's executive leadership, paving the way for a more effective use of IA within the organization and within the firm's broader Digital Transformation ambitions. Furthermore, the communication plan encompasses disseminating success stories, use cases, and the tangible benefits realized from IA initiatives, serving to educate, inspire, and galvanize the organization around the Digital Transformation agenda. This rationale extends throughout the company's Digital Transformation efforts, where the strategic implementation of IA serves as a linchpin in redefining how businesses operate, engage with customers, and innovate.

How to Craft an Executable Roadmap of IA Opportunities with Measurable Milestones for Digital Transformation

The roadmap begins with a deep dive into understanding the current state of business processes across the enterprise, identifying those ripe for automation through a lens of efficiency, cost reduction, and enhanced customer experience. These tasks are not a mere inventory of processes but a strategic assessment that prioritizes opportunities based on their potential impact on the enterprise's strategic goals, technical feasibility, and ability to enhance competitive advantage in the fast-evolving digital marketplace.

Engagement with stakeholders across the enterprise is crucial. By educating and aligning them with the IA vision, the enterprise can foster a culture of innovation where automation opportunities are readily identified and championed by those looking to improve the performance of their function or line of business. This cultural shift is foundational, turning potential resistance into proactive participation, thereby ensuring that IA initiatives have the backing and the insights needed for success.

In parallel, the roadmap could consider incorporating emerging technologies beyond IA, such as Cloud Computing, Generative AI, and Blockchain. When synergistically combined with IA, these technologies can unlock new levels of automation, intelligence, and insight, driving the enterprise towards a future where decision-making is data-driven, operations are streamlined, and customer experiences are personalized and engaging.

The delivery of IA projects ensures that the roadmap is executed flexibly, allowing for quick wins that build momentum and support for the initiative while laying the groundwork for more complex transformations. This approach ensures that projects are delivered with speed and adaptability, adapting to lessons learned and evolving enterprise needs. Throughout this journey, the focus remains steadfast on Digital Transformation's ultimate goal: redefine the enterprise's operating model by leveraging digital technologies to deliver unprecedented value.

Moreover, the roadmap is not static but a living document that evolves as new opportunities emerge, technologies advance, and the enterprise's strategic objectives evolve. It is a cycle of assessment, implementation, learning, and refinement that propels the enterprise forward, ensuring that its Digital Transformation journey is one of sustained growth, resilience, and competitive advantage.

In the end, the successful execution of this roadmap relies not just on the technologies implemented but on the enterprise's ability to foster a culture of innovation, collaboration, and continuous improvement. By focusing on strategic alignment, stakeholder engagement, and the seamless integration of emerging technologies, the enterprise can navigate the complexities of Digital Transformation.

Harmonizing Technology and Process Insight: The Journey to Becoming Digitally Native

In the **"First Wave"** of Digital Transformation, organizations should identify and seize opportunities for immediate impact through Intelligent Automation (IA). This phase is characterized by targeting low-complexity, high-volume tasks that are ripe for quick wins. For example, an insurance company might focus on automating its claims processing system, transforming a traditionally manual and time-consuming process into a streamlined, efficient operation. Similarly, tasks that were once offshored due to their rules-based, repetitive nature, such as data entry or basic customer service inquiries, are brought back onshore and automated, significantly reducing costs, lowering error rates, and improving response times.

During this initial wave, the IA program collaborates closely with process optimization initiatives, such as Lean Six Sigma projects or Process Mining, to automate, reimagine, and streamline processes. This partnership ensures that processes are not just automated in their current state but are optimized before automation, laying a solid foundation for more significant transformation efforts. This could involve simplifying the steps in a loan approval process before implementing RPA to handle applications, thus reducing the likelihood of errors and improving customer satisfaction.

The **"Second Wave"** builds on the successes and learnings from the first, expanding the scope of Digital Transformation across more complex processes and additional areas of the company. This phase involves a deeper integration of digital technologies, such as Machine Learning and NLP, to automate tasks and provide insights that drive business decisions. For instance, a bank might leverage AI to analyze customer data and offer personalized banking services, enhancing customer engagement and loyalty. This wave is more strategic, focusing on achieving a holistic transformation that encompasses operational efficiencies, new business models, and revenue streams.

The **"Third Wave"** operates on two levels. The first level involves replicating the successes of the first two waves in other parts of the organization, effectively creating a multiplier effect. Processes in areas not yet touched by automation are identified and optimized, ensuring that the benefits of Digital Transformation are felt company-wide. The second level focuses on maintaining the "Digitally Native" status achieved in the earlier waves. This involves embedding a continuous improvement and adaptability culture, ensuring that the organization can proactively respond to new technologies and market disruptions. Governance structures and technology platforms are established to support ongoing innovation and transformation, making Digital Transformation a part of business-as-usual (BAU) operations.

For example, after automating customer-facing processes and back-office operations, a company might focus on leveraging Generative AI for sentiment analysis and inference or adopting Cloud solutions for more flexible, scalable IT infrastructure. Meanwhile, areas of the business that have undergone Digital Transformation continue to evolve, embracing new technologies like IoT for enhanced product offerings or virtual reality for immersive customer experiences.

Throughout these waves, the key to success lies in the organization's ability to blend process complexity with organizational readiness. Each wave introduces new technologies and processes and prepares the organization culturally and operationally for the next level of transformation. By focusing on quick wins initially and gradually tackling more complex challenges, organizations can ensure a smooth transition to becoming Digitally Native entities capable of sustaining innovation and growth in the face of constant change.

Moving through this transformative journey, the maturation of emerging technologies is both a catalyst and a testament to the company's progress toward becoming Digitally Native. Initially deployed to address specific challenges, these technologies become intricately woven into the fabric of the enterprise. Their evolution mirrors the company's transformation, reflecting the continuous cycle of innovation and integration essential to becoming a Digitally Native enterprise.

The path to digital nativity intertwines the adoption and advancement of Intelligent Automation, emerging technologies, and comprehensive process optimization. This narrative showcases the importance of a balanced approach to Digital Transformation, where immediate technological benefits are harmonized with strategic process optimization, guiding businesses to thrive and lead in the digital economy.

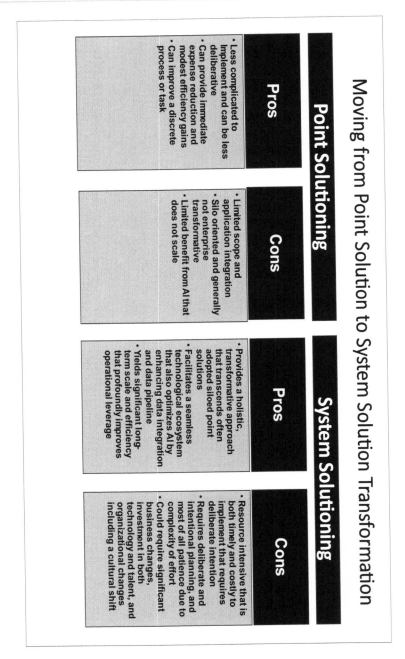

Figure 4.2 (Ajay Agrawal J. G., 2022), The Pros and Cons of Point Solutions vs. System Solutions

Case Study: A Legacy Bank's Journey to Digital Agility

When Michael was appointed as the Chief Digital Transformation Officer at Union National Bank, he stepped into a world steeped in tradition and resistant to change. The bank, while prestigious, grappled with inefficiencies that belied its storied past. High operational costs and sluggish processes were the norm, symptoms of a deep-seated aversion to altering the status quo. Michael's task was Herculean: to champion a digital renaissance that would streamline operations through Intelligent Automation and realign the bank's culture towards innovation and agility.

From the outset, Michael confronted the entrenched perception of IA as merely a tool for piecemeal solutions, quick fixes that offered temporary relief without addressing the systemic inefficiencies. This view was especially prevalent among frontline management, whose decades of service in the bank had instilled a cautious approach to operational changes. Michael recognized that he needed to demonstrate IA's potential to revolutionize isolated tasks and entire processes to shift this perspective. Through targeted workshops and carefully chosen pilot projects, he began illuminating the broader applications of IA, choosing high-impact areas where automation could deliver immediate and visible benefits. These initial successes served as proof points, slowly eroding the skepticism of frontline managers who began to witness the tangible benefits of automation.

As IA initiatives began to take root within the operational fabric of the bank, Michael's challenge evolved. The task now was to weave these disparate automation efforts into the bank's strategic framework, ensuring that Digital Transformation initiatives were not isolated endeavors but integral components of the bank's overarching business objectives. Middle management, however, emerged as gatekeepers of the established order, viewing the digital push as a threat to their domain. Michael tackled this resistance by fostering a culture of continuous learning and inclusivity, engaging middle managers in the transformation process, and emphasizing their role in the bank's digital future. He introduced process mining techniques to identify and elucidate inefficiencies, leveraging data-driven insights to build a compelling case for the strategic integration of IA.

Yet, the most formidable challenge lay ahead: securing the endorsement of the Board of Directors. Accustomed to caution, the Board was initially hesitant to sanction substantial investments in what they perceived as experimental technologies. Armed with a robust portfolio of success stories, efficiency gains, and strategic forecasts, Michael presented a compelling vision of the bank's future. He outlined a phased investment strategy, underpinned by solid metrics and early successes, to assuage the Board's concerns and align the Digital Transformation strategy with the bank's core mission of exemplary customer service.

Two years into his tenure, the transformation was palpable. Union National Bank had embraced IA across its operations and cultivated a digital-first culture that permeated every level of the organization. Processes that once lumbered along now operated with unprecedented efficiency, customer satisfaction had soared, and the bank had positioned itself as a leader in the digital banking landscape. Beyond the technological leap, Michael had ignited a cultural shift, fostering a collective belief in innovation's value and adaptability's importance.

Reflecting on the journey, Michael saw beyond the technological achievements to the broader narrative of change he had authored. Union National Bank's evolution from a legacy institution to a digitally agile enterprise stood as a testament to the transformative power of Intelligent Automation and the enduring impact of visionary leadership.

Chapter 4: Key Points Recap: Developing a Strategic Roadmap for Intelligent Automation

- **Strategic Imperative of IA Alignment:** The alignment of IA initiatives with the overarching Digital

Transformation strategy is vital. As a tactical tool within this strategic framework, IA operationalizes Digital Transformation goals on the ground, providing targeted solutions to specific organizational needs.

- **Point Solution versus System Solution**: IA typically operates as a Point Solution, addressing specific, immediate business needs through targeted automated solutions. This contrasts with the System Solution approach of Digital Transformation, which encompasses a comprehensive, integrated plan impacting the entire organizational ecosystem.
- **Practical Framework for Prioritizing IA Opportunities:** Developing a practical framework for identifying and prioritizing IA opportunities is critical. This involves assessing processes for automation readiness and potential impact, ensuring alignment with strategic business goals.
- **Effective Communication Strategies:** Establishing an effective communication plan is essential for clarifying the role and expectations of IA within the enterprise. This helps set realistic goals, educates stakeholders, and aligns IA initiatives with broader Digital Transformation efforts.
- **Executable Roadmap with Measurable Milestones**: Crafting an executable roadmap for IA with measurable milestones guides the systematic

implementation of automation projects. This roadmap should include a clear sequence of actions from understanding current processes to deploying IA solutions, ensuring they align with strategic objectives.

- **Building on Digital Transformation Waves**: The journey of Digital Transformation through IA unfolds in waves—from initial quick wins to more complex integrations. Each wave builds on the previous, expanding the scope and depth of Digital Transformation across the organization.
- **Governance and Sustainability**: Establishing robust governance frameworks and sustainable practices ensures that IA initiatives are effective and yield long-term benefits. This involves regular evaluations and adaptations to align with evolving business needs and technological advancements.

This chapter outlines the strategic roadmap for integrating Intelligent Automation within the larger context of Digital Transformation. It offers insights into overcoming resistance, aligning initiatives with strategic goals, and fostering a culture of innovation to ensure the successful transformation of a legacy enterprise into a digitally agile entity.

In Chapter 5, we will discuss the importance of having an IA governance framework to ensure sustainability for the company's Digital Transformation efforts.

Chapter 5: The Critical Role of an Intelligent Automation Framework in Digital Transformation

The Essential Components of a Comprehensive IA Framework

Establishing a robust Intelligent Automation governance framework should be seen as the foundation upon which your program will build its success. In essence, what this means for your program is how it will engage the broader enterprise to establish a pipeline of automation opportunities, determine which opportunities make sense to move forward with, and how to maintain those opportunities in production. However, governance is often seen as a burden that provides unnecessary bureaucracy that slows innovation and program velocity. While this can certainly be true if the framework is not well thought out, this should be considered the exception for that reason and not the rule (Ajay Agrawal J. G., 2022).

Merriam-Webster defines governance as "the act or process of overseeing the control and direction of something." By providing your IA program with a strong governance framework, you enable your program's ability to establish controls that will allow you to ensure that the processes within your program are moving in the right direction and are sustainable. In addition, this can be further enshrined by articulating aspects of the framework with written policies, standards, and procedures. For Intelligent Automation, six critical pillars of your governance framework should be advanced: Engagement, Delivery, Development, Governance, Infrastructure, and Operations.

Engagement

It is essential to think of this pillar as establishing your relationship management function between the program and the broader enterprise. In doing so, you will want to create a communication plan which will lay out your engagement efforts. An important part of this plan should be focused on educating the enterprise on what IA is and what IA is not and on how IA differs from traditional software development.

This will pay dividends in helping the business find opportunities and level-set expectations concerning IA's bot journey or lifecycle (from opportunity identification to deployment). It will also provide exposure to the enterprise on the promise and value proposition of IA, align the enterprise to what is needed to succeed with an IA program, and establish a working relationship with senior leaders throughout the enterprise regarding automation and transformation.

Lastly, one of the final activities on a typical bot journey is called Benefit Realization. This is typically done once the bot is deployed for 6 to 12 months to ensure that the assumptions made for the automation have materialized for the critical metrics determined for your program. Reporting on both forecasted and realized benefits should be a regular and reoccurring expectation from this pillar. The ultimate goal of all of the engagement activities is to help build a robust pipeline of automation-ready IA opportunities (stable, standardized, and repetitive processes) that are proactively primed by a highly IA-educated and motivated enterprise.

Delivery

The Delivery pillar establishes how you will define and iterate through your bot journey or lifecycle from the current process understanding and documentation phase (As-Is) to the automation solution design (To-Be) and then the development to deployment phases in a continuous integration/continuous (CICD) delivery fashion. For IA delivery, most organizations develop and deploy using Agile. In doing so, you will have your typical Scrum Master, QA Tester, and IA Developers. However, two roles on your delivery teams are critical to building robust IA solutions: the IA Business Analyst and the IA Solution Architect (or Tech Lead by some).

Also, typically, within IA delivery, your IA Business Analyst (BA) is "quarterbacking" in the earliest phases of automating the process and providing critical support in the later phases during development and deployment. In the earlier phases, the BA is working to document the process, determine its feasibility, conduct walkthrough sessions with the subject matter experts to understand and document the As-Is process, and determine the solution design, working hand in hand with the Tech Lead throughout these phases.

From there, the BA will use the solution design along with additional technical input provided by the Tech Lead to write stories for the backlog for development purposes. Lastly, the nature of most IA projects is that they are relatively small, short-term, and discrete initiatives, and given this, your project management approach should reflect this.

Proposed IA COE Delivery Team – Team Structure and Roles

- **IA COE Engagement:** Facilitates collaboration between the COE and business functions for automation opportunities, ensures processes are ready for automation, promotes IA education and engagement enterprisewide, and aids in tracking business outcomes.
- **IA Product Owner**: Identifies business opportunities, manages the backlog, coordinates with application teams for access, resolves impediments, and is responsible for outcome tracking with Line of Business (LOB) support.
- **IA Business Analyst (BA):** Assesses feasibility and potential benefits with business and finance teams, document current processes, collaborates with the Solution Architect on solution design, testing, and user stories, and oversees automation from inception to deployment following IA Development Lifecycle.
- **IA Scrum Master:** This person undertakes standard scrum master duties and oversees Agile practices, metrics, and deployment activities.
- **IA Solution Architect:** Partners with BA on designing automation solutions and user stories, maintains documentation, develops complex automation requirements, explores new capabilities through POCs, and defines development governance.
- **Developers:** Responsible for IA development and engages in special projects as directed by the Solution Architect.
- **QA Testers** Conduct tests based on defined criteria and scenarios in conjunction with the Solution Architect, ensuring clarity in testing documents.

Figure 5.1, An example of an IA Delivery Team Structure and Roles

Development

While development does happen within the Delivery pillar, it should be thought about independently as a car manufacturer thinks about engineering. The reason for this is that many adopters of IA (or, in its rawest form, RPA) often complain about how fragile automations are in production, the frequency in which they fail, and their low quality overall. While one part of this must be covered in the delivery pillar in ensuring that the exception paths are well documented from the current manual process and how robust the solution design is architected, a large part must be addressed in how the automation is engineered and developed. It is essential to proactively create a set of standards based on the development best practices utilized for the IA platform your firm decides to employ.

These development standards should be enforced using a formalized code review process that will ensure consistency of adoption of your minimal development requirements. In many cases, the failure of an automation has as much to do with the development approaches taken in building it as any other factor. The outcome you want from this pillar is to ensure that the automations developed by your practice are robust, stateful, and resilient in production, and able to deal with minimal production environmental change without failing. Suppose there are issues with the automation in production. In that case, you want to ensure that the automation is engineered in such a way that issues can be quickly identified and remediated.

Governance

This is often a misunderstood and undervalued function within most organizational make-ups, let alone an expressed strategic pillar. This is a big mistake because a strong governance function can drive uniformity and consistency within your program and ensure that your program is ready for any audit or regulatory reviews that are common within industries such as Financial Services. In addition, your governance function should assist the program in understanding the risks that the program poses to itself and the enterprise as a whole and determine what controls should be in place to mitigate those risks.

Another central responsibility that Governance should have is formalizing your program's policies, standards, and procedures. By doing so, you are adding the authority of compliance to those activities that you feel should be mandated to the entire enterprise as "musts" and will have the assurance that audits will be conducted to guarantee those "musts" are complied with. The reason for this is to safeguard that regardless of whether you are federated or centralized if your program is the central authority for IA within the enterprise, you need to warrant that you have some clear rules for the road.

Finally, your governance function could help with things such as role alignment to the framework and bot journey as a part of creating and maintaining a "Responsible, Accountable, Consulted, and Informed" (RACI) matrix or with things like helping to draft needed RFPs for the program or reviewing SOWs working with teams like Sourcing. Still, these are not necessary core requirements for the pillar.

Infrastructure

This pillar primarily deals with the initial implementation and periodic upgrades of the Intelligent Automation platform that your firm decides is right for its needs. This means determining the architectural design for your IA infrastructure and deciding what hardware will be needed, how it will be laid out to support this effort from a diagram standpoint, and what scheme you will need to host your VDIs so that your digital workers will have a place to do their work. This will also mean devising a disaster recovery plan, which might entail rolling your IA infrastructure from a primary site to a secondary fallback site in case of natural disasters, general failures, or other matters that could cause your primary site to fail.

It is important to remember that not having a solid disaster recovery plan means that your entire digital workforce will be at risk, including those running critical processes for your business that you have decided to automate. Further, this pillar will have information security responsibilities and ensure that your IA platform and the automations that will run from it are adequately secured, logged, and vaulted. Finally, this can also involve other auxiliary technologies like Process Mining or Intelligent Document Processing (IDP) that will support your Intelligent Automation efforts with additional capabilities.

Operations

Once your automation is deployed, your Operations or Production Support team ensures that your digital worker "shows up for work" based upon a plan devised by its process owner per its solution design. However, your IA program and process owners must understand some crucial notions once their automation is in production. Firstly, the IA Operations team will monitor the automation to ensure it is running as conceived and report any issues with the automation to the process owner and other stakeholders. Secondly, the IA program does not own the process, nor the automation performing the process. The reason for these two notions is that accountability for the process should not shift from the business to the IA program. For example, if they were managing a human worker, the business should be actively monitoring that worker's output to ensure that it is accurate and as expected.

Lastly, the responsibility for credentials management for the digital workforce should be maintained and managed by your Operations team. The reason for this is that depending on the organization, managing credentials for the digital workforce can be quite complex given the array of credential types, from Single-Sign-On (SSO), Non-Personal Service Accounts to Third-Party credentials for applications and systems utilized outside of the company's tech stack. Some organizations will choose to mirror the human workforce for credentials, while others will use some hybrid approach. However, given the various credential types, varying renewal schemes must be centrally managed.

Again, as discussed in the previous chapter, an important concept to keep in mind is that Intelligent Automation is the use of technology to automate a manual process across the computer applications and systems utilized by that process in a deterministic or rules-based way. Given that you are not building an application, platform, or system, IA should be viewed through a different set of lenses than traditional software development. Many IA programs fail because they are not deliberate in distinguishing the difference between these seminal concepts. What the pillars provide you is a sustainable structure where you can bucket all of your key activities and roles to organize and operate your program (your IA factory) (Marco Iansiti K. R., 2020). A solid governance framework will enable you to build a high-quality "Enterprise Automation" in a consistent, predictable, and repeatable fashion that is both inherently scalable and maintainable as it moves through your

IA factory, which should align with your Digital Transformation strategy. Encapsulating your program within an Intelligent Automation Center of Excellence (CoE) is the natural next step in galvanizing the foundation of your IA program. If established robustly, a strong IA CoE will be the conductor of your IA governance framework and a cornerstone to your IA program's success and your company's Digital Transformation efforts as a whole.

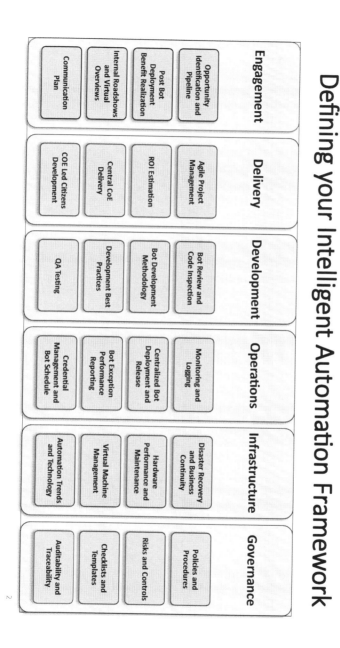

Figure 5.2, An Example of a Pillar-Based IA Governance Framework

The Strategic Value of Establishing a Dedicated IA Center of Excellence (CoE)

Establishing an IA CoE is a strategic move that amplifies the benefits of IA by centralizing expertise, standardizing practices, and fostering a culture of continuous improvement and innovation. A well-structured IA CoE serves as the nucleus of IA initiatives, providing the necessary governance, methodologies, and tools to ensure that automation projects align with the company's strategic goals and Digital Transformation efforts. This centralized approach ensures a cohesive and coordinated effort toward Digital Transformation, steering clear of the pitfalls associated with fragmented and siloed automation projects.

One of the core advantages of an IA CoE is its ability to facilitate a comprehensive and holistic view of automation opportunities across the organization. By systematically identifying, assessing, and prioritizing automation projects, the CoE ensures that the most value-generating and strategic processes are automated first. This strategic prioritization not only optimizes resource allocation but also maximizes the impact of automation on business performance.

Moreover, an IA CoE plays a crucial role in cultivating an Enterprise Automation Development Mindset. This mindset shifts the perception of automation from a series of isolated projects to an integral part of the organizational fabric, akin to any other critical enterprise technology (moving from Point Solution to System Solution thinking). Adopting this mindset necessitates a strategic vision that aligns automation efforts with the broader business objectives, ensuring that each automated process contributes to efficiency, scalability, and innovation.

Furthermore, the integration of IA within the Digital Transformation strategy fosters a culture of innovation and enables businesses to stay ahead in this competitive landscape. The agility afforded by IA allows businesses to adjust to market changes and customer needs quickly, providing a competitive edge in rapidly evolving industries.

The strategic value of establishing a dedicated IA CoE lies in its ability to orchestrate and accelerate Digital Transformation initiatives. Through centralized governance, standardized methodologies, and a holistic approach to automation, IA CoEs empower businesses to realize their Digital Transformation objectives effectively. By embedding an Enterprise Automation Development Mindset and aligning automation efforts with strategic business goals, organizations can harness the full potential of IA to drive operational excellence, foster innovation, and secure a dominant position in this time of rapid technological change. This explanation underscores the symbiotic relationship between IA CoEs and Digital Transformation strategies.

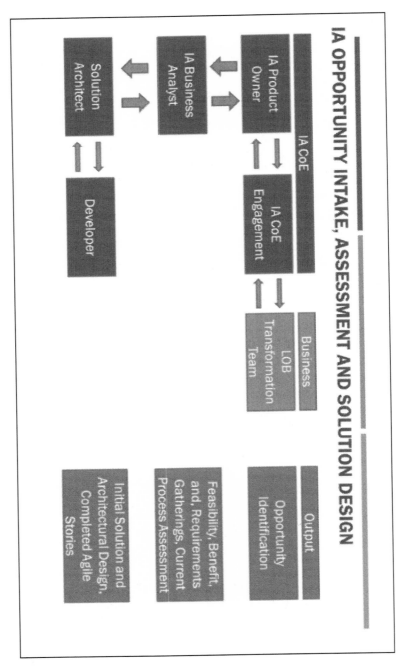

Figure 5.3, The output of certain critical roles of an IA CoE

The Guidance on Adapting and Customizing IA Frameworks to Organizational Maturity Levels

The advent and proliferation of Intelligent Automation within the business landscape signify a monumental shift towards optimizing efficiency, enhancing customer experience, and ultimately securing a competitive advantage in the rapidly evolving digital economy. This transformation is not merely about integrating new technologies but about redefining the very essence of organizational operations, culture, and strategic direction. As companies navigate this journey, the customization and adaptation of IA frameworks to align with individual and organizational needs and maturity levels emerge as a cornerstone for success.

Adopting and integrating IA into an organization's fabric is multifaceted and will require a nuanced approach that considers each entity's unique challenges, objectives, and cultural dynamics. As such, guidance on customizing IA frameworks becomes indispensable, providing a blueprint for organizations at various stages of their Digital Transformation journey.

For a company just embarking on its IA journey, the inception of an IA Center of Excellence (CoE) from scratch presents a unique set of challenges and opportunities. At this nascent stage, the focus should be on laying a solid foundation that encompasses the technical aspects of IA and the cultural and strategic dimensions. It involves a meticulous process of understanding the current state, envisioning the desired future state, and meticulously planning the trajectory to bridge this gap. Essential steps include educating the workforce on the potentials and limitations of IA, developing a strategic roadmap aligned with business objectives, and fostering a culture of innovation and continuous improvement.

For companies restarting their IA CoE due to previous leadership or knowledge gaps, the approach entails a careful analysis of past efforts to identify pitfalls and glean lessons learned. This scenario demands a robust governance framework to ensure alignment with organizational goals and the establishment of clear accountability mechanisms. Restarting an IA CoE offers a unique opportunity to reset expectations, refine strategies, and leverage emerging technologies more effectively. This involves re-engaging stakeholders across the organization, reassessing the automation pipeline, and placing emphasis on building a resilient and adaptable CoE that creates robust and sustainable automated solutions.

For a maturing IA CoE looking to advance its Digital Transformation journey, the challenge lies in scaling successes, integrating advanced technologies, and fostering an innovation-driven culture. This stage is characterized by the expansion of IA's reach across the entire enterprise, leveraging insights from data analytics to drive strategic decisions and continuously optimizing processes for greater efficiency and effectiveness. A maturing CoE must focus on sustainability and being a robust part of the company's digital native equation. It seeks ways to embed IA into the organization's Digital Transformation ethos and explore opportunities to innovate and create value beyond operational efficiencies, such as helping to decrease the overall risk of the firm via reduced error rates or removing the possibility of bad actors through digital automation, or by helping to drive revenue by increasing the scalability of specific processes that are accretive to the company's top line.

The path to Digital Transformation through IA is a continuous journey shaped by the evolving landscape of technologies and the dynamic needs of businesses. As organizations traverse this path, the adaptation and customization of IA frameworks to their unique contexts underscore the importance of a strategic, holistic approach. This journey, while challenging, offers unparalleled opportunities to redefine the future of work, foster innovation, and achieve sustainable competitive advantage.

Guidance on Measuring Success for IA

The guidance on measuring success for IA in terms of ROI, lowering risk, enhancing revenue, lowering overall expenses, and avoiding costs is intrinsically linked to the broader goal of Digital Transformation. As businesses integrate IA into their operations, adopting a multi-faceted approach to determine the effectiveness and efficiency of these initiatives becomes imperative.

IA's success lies in its capacity to deliver tangible, measurable outcomes. Return on Investment (ROI) is a paramount metric calculated by comparing the financial gains from IA against the costs incurred to implement such solutions. Financial gains often manifest through reduced operational costs, improved productivity, and avoidance of potential costs associated with manual processes prone to errors. Cost avoidance, another crucial measure, captures the savings from preempting manual errors, delays, and the utilization of human resources in mundane tasks. Through meticulous documentation and analysis, businesses can quantify the number of hours saved, errors reduced, and the overall speed enhancement in processes, subsequently converting these figures into financial terms.

Risk mitigation is quantifiable by reducing compliance violations, security breaches, and operational risks, directly correlating with financial savings from avoided regulatory penalties, fines, and operational losses.

Although more complex to measure, enhancing revenue through IA can be seen through increased sales volumes, higher customer retention rates, and the creation of new revenue streams enabled by IA-driven innovations. Each of these quantitative measures requires a robust data collection, analysis, and reporting framework, ensuring stakeholders can clearly see the financial impact of IA initiatives.

Beyond numbers, the success of IA also deeply impacts the qualitative aspects of business operations. Improving customer and employee experience stands out as a critical qualitative measure. IA initiatives often lead to faster response times, higher accuracy in customer service, and personalized customer interactions, enhancing customer satisfaction and loyalty. Automating tedious tasks allows employees to focus on higher-value activities, leading to better utilization of human resources. This typically leads to increased employee satisfaction while improving process efficiency, efficacy, and effectiveness.

Though difficult to quantify, organizational agility and resilience are essential qualitative outcomes. IA equips businesses with the flexibility to adapt to market changes rapidly and maintain operations amid disruptions, a competitive advantage that, while qualitative, has substantial long-term benefits.

Example Use Cases Quantifying the Potential of IA

- **Financial Services**: A financial services firm integrates IA to automate its loan processing system. The quantitative benefits could include a 60% reduction in processing time and a 40% decrease in operational costs. Qualitatively, the firm will experience enhanced customer satisfaction due to faster loan approvals and significantly reduced human errors, leading to improved compliance and risk mitigation.
- **Healthcare:** A healthcare provider employs IA to manage patient records and appointment scheduling. Quantitatively, this might result in a 50% reduction in administrative costs and a 60% decrease in appointment no-shows. Qualitatively, patient experience will markedly improve through timely communications and personalized care coordination, while healthcare professionals report higher job satisfaction due to reduced administrative burdens.
- **Retail:** A company uses IA for inventory management and customer service chatbots in the retail sector. The quantitative outcomes could include a 25% decrease in inventory costs and a 35% increase in online sales. Qualitatively, the company will benefit from an enhanced online shopping experience, with customers

expressing high satisfaction with the instant support provided by attended automation-powered chatbots, leading to increased loyalty and brand reputation.

Measuring the success of Intelligent Automation within the context of Digital Transformation requires a balanced approach that considers both quantitative and qualitative metrics. Businesses can gain a comprehensive understanding of IA's value by meticulously evaluating the financial impacts, risk reduction, revenue enhancement, and cost avoidance, alongside the improvements in customer and employee experiences, organizational agility, and operational resilience. Through strategic implementation and ongoing evaluation, Intelligent Automation achieves specific operational goals and propels the organization forward in its Digital Transformation journey, ensuring long-term competitiveness and sustainability.

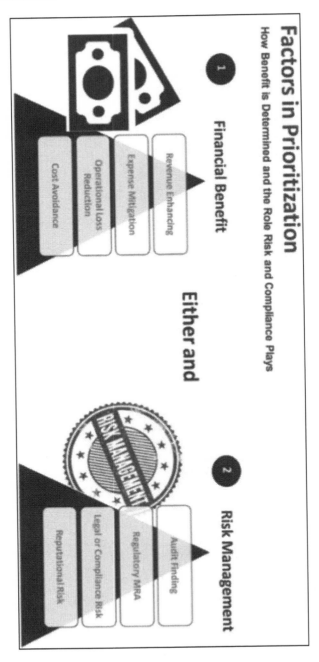

Figure 5.2. The Two Mutually Important Criteria Sets for IA Prioritization

Addressing Organizational Resistance to Intelligent Automation

Businesses often face the dual challenge of innovating rapidly while ensuring their workforce transitions smoothly into new digital paradigms. This journey is about integrating new technologies and fundamentally reimagining how organizations operate, engage with customers, and empower their employees. As companies embark on this transformative path, they encounter various obstacles, including organizational resistance, the need for effective change management, concerns over job displacement due to automation, and the essential role of leadership in steering the organization toward a harmonious integration of human and digital capabilities.

Organizational resistance often stems from a fear of the unknown and a comfort with the status quo. This resistance is not just a hurdle but an opportunity for leadership to engage, educate, and enlist the workforce in the transformation journey. Successful Digital Transformation initiatives go beyond merely adopting technologies; they require a cultural shift toward embracing change, experimentation, and continuous learning. Leadership can play a critical role in this process, setting a vision that aligns Digital Transformation with the organization's core values and strategic objectives. By fostering a culture of transparency, open communication, and inclusivity, leaders can demystify the transformation process, highlighting the 'what' and 'how' and, more importantly, the 'why' behind the change.

The fear of job displacement due to Intelligent Automation is a significant concern among employees. This fear can lead to increased resistance and decreased morale if not appropriately addressed. Organizations must recognize that while some roles may become automated, new opportunities for more meaningful and impactful work will arise. The discussion should shift from job replacement to job enhancement and creation. By offering retraining and reskilling programs, organizations can prepare their human workforce for future jobs where digital tools and technologies complement human creativity, strategic thinking, and emotional intelligence.

As organizations transition to digitally enhanced operations, employees' roles have to evolve. This shift necessitates a reimagining of job roles and responsibilities. Employees will need to adopt a mindset of lifelong learning, continuously adapting to new tools and methodologies. In this context, digital literacy becomes fundamental, and the ability to work alongside digital colleagues and the understanding of their capabilities and limitations becomes a critical skill.

Leadership's role in quelling misgivings and fears about Digital Transformation is multifaceted. Leaders must act as champions of change, embodying the digital-first mindset they wish to instill in their organization. They must ensure the transformation journey is inclusive, providing clear pathways for every employee to contribute to and benefit from the change. This includes transparent communication about the impacts of automation, investment in training and development programs, and creating a supportive environment where experimentation and failure are seen as stepping stones to innovation.

The successful integration of Intelligent Automation within a digitally transformed organization requires a holistic approach that addresses both technological and human factors. It demands visionary leadership, a culture that embraces change, effective change management practices, and a commitment to nurturing the human workforce's growth and adaptation. By doing so, organizations can achieve operational efficiencies and competitive advantages and foster a more engaged, empowered, and fulfilled workforce.

Case Study: Rising from the Ashes of IA Program Attempts

In the fast-paced world of telecommunications, how Star Telecom, under the leadership of CIO Roger Telephony, eventually established a successful IA CoE presents a compelling narrative of resilience, strategic foresight, and the transformative power of Intelligent Automation (IA).

Under the stewardship of Sergio NoClue, Star Telecom's initial venture into Robotic Process Automation (RPA) was marked by enthusiasm yet hampered by a lack of strategic direction. This period, characterized by exploratory projects and a limited understanding of automation's potential, saw RPA as more of a novelty than a strategic tool. The projects lacked a cohesive vision, failing to bridge the chasm between technological ability and business needs. While pioneering, this approach underscored the necessity of aligning automation with overarching business goals, a lesson that would become pivotal in the journey ahead.

Recognizing the need for a more structured approach, Star Telecom's leadership embarked on a second attempt to harness the power of automation. This phase was marked by a strategic recognition of automation's potential to transform operations and enhance competitiveness. Despite this strategic intent, the program stumbled due to inadequate governance, lack of development standards, and insufficient talent and leadership to build enterprise-worthy automations for the company's critical processes. This phase highlighted the vital importance of a robust governance framework, a comprehensive opportunity pipeline process, and standards for development and monitoring, as detailed in the discourse on establishing robust governance frameworks for IA. Learning from the past, Roger Telephony presented a compelling case to the CEO and the board for a renewed foray into IA. Drawing on the lessons learned from previous attempts, he outlined a vision grounded in strategic alignment, robust governance, and the nurturing of talent. This vision was underpinned by the recognition that Digital Transformation and IA are not merely about technology but about reimagining business processes and culture in ways that drive sustainable competitive advantage.

Telephony's pitch for the IA program's restart was not just a call for investment but a manifesto for change. It emphasized the need for a holistic approach to automation that integrates IA into the organization's fabric, supported by a governance framework that ensures alignment with business objectives. He highlighted the critical role of leadership and the strategic imperative to build a talent pool equipped to drive the program toward success.

As Star Telecom stands on the brink of this new chapter, the focus is on embedding IA within a strategic framework that addresses past oversights. This entails developing a governance structure that oversees the implementation of automation solutions and ensures they are leveraged to propel the company toward its long-term objectives. It involves cultivating an enterprise-wide mindset that sees automation as a strategic enabler supported by continuous learning and adaptation.

However, this also includes getting support for IA from other critical pillars of IT such as Identity Access Management (IAM), Cybersecurity, and IT related Risk Management, along with ensuring the application teams throughout the enterprise ensure that their change management procedures factor in software robotics as a possible constituent of their platform. While unique and distinct from a need standpoint, Star Telecom had to come to contemplate the factors that would allow its digital workforce to be successful, just as it had to do for its human workforce.

In this story, Roger Telephony emerges not just as a technologist but as a visionary leader who understands that the journey toward effective automation is iterative and learning-driven. He recognizes that success lies in adapting, learning from past challenges, and relentlessly pursuing a vision that aligns technology with strategic business outcomes. As Star Telecom embarks on this renewed journey, governance, strategy, and leadership lessons serve as guiding lights, promising a future where IA becomes a cornerstone of innovation and competitive advantage.

Chapter 5 Key Points Recap: The Critical Role of an Intelligent Automation Framework in Digital Transformation

- **Establishing a Robust IA Governance Framework:** A strong governance framework lies at the heart of a successful IA program. It encompasses policies, standards, and procedures, along with comprehensive controls to ensure IA initiatives are aligned with organizational goals and sustainable over time. This governance framework is crucial for navigating the complexities of Digital Transformation and maximizing IA's strategic value.
- **Engagement as a Key Pillar:** Effective engagement strategies are essential for fostering a collaborative environment where IA is embraced across the enterprise. By educating stakeholders on IA's benefits and nuances, such as how developing an automated process differs from traditional application development, organizations can cultivate a culture that actively seeks automation opportunities and supports IA's lifecycle from inception to benefit realization more effectively.

- **Strategic Delivery and Development Processes:** The delivery and development phases of IA projects are critical for ensuring that automations are solution-designed, architected, and developed to be robust, resilient, and sustainable in production. Emphasizing Agile methodologies, these phases involve critical roles such as IA Business Analysts and Solution Architects to ensure automation initiatives are delivered efficiently to meet the enterprise's quality and scalability requirements.

- **Infrastructure and Operations Support:** A reliable and secure IA infrastructure is fundamental for deploying, monitoring, and maintaining automation solutions. This includes platform availability, bot credential management, disaster recovery planning, and information security controls to protect the digital workforce and ensure it can deliver value for the enterprise in production.

- **The Strategic Value of an IA Center of Excellence (CoE):** Establishing a dedicated IA CoE amplifies the benefits of IA by centralizing governance, expertise, development practices, checklists, templates, and artifacts to foster a culture of continuous improvement. The CoE plays a crucial role in aligning IA initiatives with business strategies and driving innovation across the organization.

- **Adapting IA Frameworks to Organizational Maturity:** Tailoring IA frameworks to fit the organization's digital maturity level is essential for effective implementation. Whether starting, restarting, or advancing an IA program, understanding where you are in your IA and Digital Transformation journey and being intentional and deliberate about getting to your digital maturity destination (goals and objectives) are key to success.

- **Comprehensive Measurement of Success:** Assessing IA's impact involves quantitative metrics, such as ROI and cost savings, and qualitative outcomes, including improved customer and employee experiences. A balanced approach to measuring success is critical for demonstrating IA's contribution to Digital Transformation and securing ongoing support for IA initiatives.

- **Overcoming Organizational Resistance:** Leadership plays a pivotal role in addressing resistance to Digital Transformation. Leaders can guide their organizations through the challenges of adopting IA by promoting transparency, fostering a culture of innovation, and investing in workforce development.

- **Star Telecom - A Case Study in Resilience and Strategic Foresight**: Star Telecom's journey under CIO Roger Telephony illustrates the significance of learning from past automation endeavors. Star Telecom overcame earlier setbacks by instituting a robust governance framework, prioritizing strategic alignment, and fostering talent development. This narrative reinforces the importance of a comprehensive IA framework and establishing an IA CoE in achieving Digital Transformation success, showcasing the transformative power of perseverance, strategic planning, and leadership in navigating the challenges of IA integration.

In summary, Chapter 5 highlights the crucial elements of an IA framework that supports Digital Transformation, emphasizing the importance of governance, strategic alignment, and the establishment of a CoE. Through careful planning, engagement, and leadership, organizations can harness the power of IA to achieve operational excellence, foster innovation, and maintain a competitive edge on their journey to becoming Digitally Native.

In Chapter 6, we will discuss the importance of having an Enterprise Automation Development Mindset and why it is critical for giving the organization confidence that automation produced by your IA program can provide the sustainable value needed by the business in terms of strategic alignment and reliability

Chapter 6: Fostering an Enterprise-Wide Automation Development Mindset

The Importance of Having an Enterprise Automation Development Mindset for Intelligent Automation

The development of enterprise-worthy IA solutions requires the implementation of robust development standards. These standards should establish clear guidelines and controls for developing, deploying, and maintaining automation solutions, ensuring that they are built to last and can deliver the value and dependency required for the process the business wants to automate. It also involves rigorous testing and quality assurance procedures to identify and address potential vulnerabilities before any inherent issues with the automation can negatively impact the operations of the business process the software robot was built to automate.

The limitations of IA become evident in scenarios involving mission-critical, long-running, and complex processes where an Enterprise Automation Development Mindset needs to take hold to ensure that the automated solution produced has the resilience, durability, and serviceability required to support effectively the solution in production. This will also better ensure that client satisfaction with the solution stays high and that the solution can maximize its ability to provide value rather than constantly needing to be repaired and redeployed.

The Common Complaint about Intelligent Automation

A discussion point often broached by many attendees of IA conferences is "Why do IA bots break down so much?" Unfortunately, the finished automated solution that many IA programs produce might not provide the resilience needed or might not have considered the high degree of application and infrastructure variability that most companies failed to control appropriately or could not control at all if the process utilized third-party systems or URLs. Because of this, IA bots were easily susceptible to subtle application changes involving the GUIs of the underlying application that a flesh and blood person would quickly adapt to. In addition, their program had not developed enterprise development standards that could lay the groundwork for their bots to scale or have the error mitigation resilience to handle the company's more critical processes, and that would take into account a great deal of this possible variability. Without accounting for this variability, IA solutions will lead to less efficient and effective outcomes while being more costly to host, maintain, or repair in production.

The Long-Term, "Temporary", Solution

Contrary to many people's perceptions of IA, it's often not an ephemeral solution meant to be a bridge to what is seen to be firmer system integration. On the contrary, IA is often the solution for the foreseeable future as it can achieve data integration and complete tasks in a human-like way deterministically with identical deliverables, if not more, that could be generated from the manual process being automated, and all without changing or altering the underlying applications being automated against.

More often than not, that Application Programmable Integration (API) project, in conjunction with the app team to build out new functionality and interfaces for the application, perpetually gets delayed, if ever completed at all. In addition, some tasks go beyond the system integration provided by solutions like APIs alone and are best suited for the multi-modal approach that Intelligent Automation delivers. They need to be architected with this intention from the beginning. Think of a solution that integrates data from documents using an Intelligent Document Processing (IDP) platform upstream and then uses IA to disseminate this information in the downstream systems used by the process via API endpoints from both the IDP platform into the downstream applications and then generates a set of tangential dashboards or other outputs, via platforms like Power BI, Tableau, or just Microsoft Excel, using that same data.

This non-acknowledgment is the problem at the heart of why firms can sour about Intelligent Automation. This is because IA can often automate processes rather quickly if you take a "happy path" approach, which allows you to get something that "works" briefly in production but does not have any of the precautions that will enable the automation to be sustainable. This inherent unreliability often undermines the initial triumph of deploying the automation in the first place, with that confidence of this initial deployment undermined if the IA solution has ongoing issues (sugar high, then disillusionment).

The Importance of Having an Enterprise Automation Development Mindset for IA

The first step towards successful Enterprise Automation is embracing an Enterprise Automation Development Mindset. This approach involves viewing automation not as a series of isolated projects that are discretely done with no reorientation or controls to development best practices but as an integral part of the organizational fabric that must be supported like any other enterprise technology. It demands a strategic vision that aligns automation efforts with overall business goals, ensuring that every automated process contributes to the broader objectives of efficiency, scalability, and innovation.

This mindset advocates for a holistic approach to automation. It encourages looking beyond simple task automation to understand and optimize end-to-end processes. This involves a comprehensive view of the organization's processes, identifying interdependencies, and ensuring that automation solutions have what's needed to support some of the enterprise's most critical processes.

An Enterprise Automation Development Mindset focuses on building scalable and sustainable automation solutions. Scalability ensures that automated processes can adapt to increased workloads or changing business needs without significant rework. Sustainability involves designing maintainable automation solutions over time, with minimal disruption and consistent performance.

Start with Good Process Understanding and Discovery for Intelligent Automation

Good Process Understanding and Discovery is a fundamental principle for Intelligent Automation (IA), emphasizing the need for a thorough understanding of the business processes and the applications involved in these processes before any development begins. This principle involves delving deep into how processes function, where they can be optimized, and how the applications integral to these processes operate.

The journey begins with detailed process mapping, documenting each step, information flow, and decision points. This mapping creates a blueprint of the current operations, highlighting areas where automation can be most beneficial. This also includes understanding the applications used at each step, ensuring that any automation is compatible and enhances existing systems. This principle is critical to identifying bottlenecks, redundancies, and inefficiencies in the current process and how the applications are used. It is imperative to understand that application details are as important as understanding the process steps' requirements.

Processes often vary across different scenarios, as do the applications' capabilities. A thorough discovery process will reveal these variations and the flexibility or limitations of the applications involved. This understanding is crucial for designing an automation solution that is adaptable and can handle the nuances of real-world operations. Engagement with process owners, subject matter experts, and other stakeholders is vital. This engagement helps uncover not just the process details but also the user experiences, pain points, and potential areas of improvement related to application usage. In addition, data analysis is critical to understanding process performance and application efficiency. Analyzing data from both processes and applications can offer insights into how they interact, where delays or errors occur, and how automation can optimize this interplay.

Lastly, a thorough understanding of the applications used in the process directly influences the design of the automated solution. It ensures that the automation is not just layered on top of existing processes but is integrated seamlessly, exploiting the applications' capabilities and the process's efficiency.

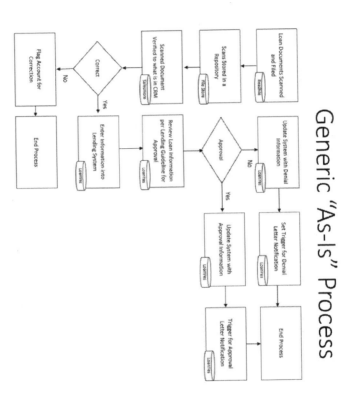

Figure 6.1, Example of a Current, or "As-Is" process

Effective Solution and Architectural Design for Intelligent Automation

When approaching Intelligent Automation (IA), the solution's design and architecture (its future state) are critical. This involves a meticulous approach that considers current needs and anticipates future scalability, integration capabilities, and technical limitations. Effective solution and architectural design in the context of IA requires a blend of technical acumen and strategic foresight.

By optimizing the IA solution for the specific applications involved, organizations can reap several benefits, including:

- **Improved Automation Efficiency:** Leverage application-specific logic, features, and data structures to optimize automation workflows and performance.
- **Reduced Integration Complexity:** Use APIs and direct database querying capabilities to minimize integration efforts and reduce bot breakdown due to changes in the GUI or any other underlying application issue.
- **Increased Data Accuracy:** Ensure data integrity and consistency by leveraging application-specific data validation mechanisms at the field level.
- **Enhanced Automation Error-Handling:** Enhance error handling of the automation workflows based on

real-time events, data specifications, and other issues that could undermine the effectiveness of the automation.

Leveraging application features and backend integration capabilities are crucial for successful IA implementation. Organizations can adopt these strategies to create robust, scalable, and application change-resilient automation solutions that deliver long-term value and drive business growth.

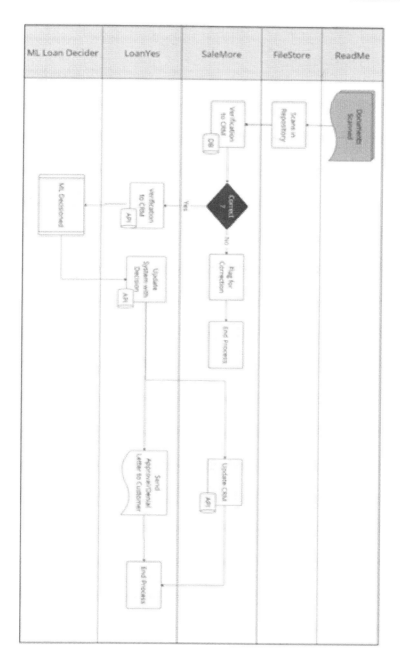

Figure 6.2, Depicts a reimagined "To-Be" process that is streamlined with the use of both RPA and ML from the manual "As-Is" process.

Establishing Standards and Governance to Uphold Principles

The adoption of Enterprise Automation principles is only the first step. Developing standards that adhere to these principles and establishing governance mechanisms to enforce compliance is crucial to ensure long-lasting effectiveness. This involves creating guidelines, procedures, and code reviews that align with principles of reliability, stability, gracefulness, and maintainability. Governance ensures that all automation projects across the organization consistently apply these standards, leading to uniform quality and performance. Such an institutional framework maintains the integrity of automation initiatives and fosters a culture of continuous improvement and strategic alignment.

- **Robust Automation Resilience and Error-Handling:** Ensure automated processes are reliable, can handle varying loads and conditions, and are not as easily subject to failure due to things like subtle changes to the application GUIs or architecture, or if there is infrastructure latency, etc.
- **Enhanced Stability and Data Scalability:** Design bots to operate consistently under different scenarios and handle processing volumes that may be materially higher than when the bot was deployed, including unexpected spikes.

- **Ensured Gracefulness:** Implement error handling and recovery mechanisms that allow bots to fail gracefully, ensuring minimal impact on operations.
- **Effective System and Application Integration:** When available, leverage backend access for deeper integration with systems and applications for better overall stability and sustainability, such as APIs and direct querying of the application's database.
- **Engrained Maintainability and Code Scalability:** Build bots that are modular and scalable, allowing for easy updates, enhancements, and annotation of the automation's code.
- **Essential Management of Technical Debt:** Proactively addressing and regularly managing technical debt is critical to maintaining the integrity, stability, and strategic alignment of Intelligent Automation solutions, ensuring they adhere to core principles and remain effective over the long term.
- **Vigorous Code Testing:** Create specific guidelines for testing automation solutions, covering various scenarios and use cases, and ensure granular unit testing.

In addition, implementing automated testing mechanisms to check the performance and functionality of automation solutions can significantly increase the effectiveness of your IA development efforts. This practice helps with the early detection of issues as the automated solution is being developed and can speed overall development through the QA and UAT phases, saving considerable time and effort for the solution deployed. Automated Testing can also help in production to allow for quick and regular regression testing of the IA solution in anticipation of an application or infrastructure change, or if the automation were to fail, provide nearly instantaneous feedback on where and why it is failing from the code, which leads to shorter periods of the bot being out of production.

A Place for Citizen Development within the Enterprise Automation Development Mindset

Citizen Development (CD) aims to empower non-technical users to create automations without the need for extensive coding expertise. While CD can democratize automation, it often lacks the rigor and discipline of traditional development practices. This inevitability leads to an automation that violates these principles and reinforces the negative perceptions of IA. By adopting the previously mentioned principles as minimum enterprise standards, CD can be aligned with the goals of Enterprise Automation. Organizations can provide guidance and training to ensure that automations are built with long-term sustainability in mind.

Lastly, adopting Enterprise Automation principles can significantly lower risk management for Citizen Development by providing a structured and disciplined approach to automation development and deployment and eliminating commonly viewed bad practices such as hard-coding credentials and lack of variables where there was the most sensible option in the automation code.

It can also lower the risk to applications associated with the process, especially those associated with processes where data accuracy in those systems is paramount. In addition, it significantly reduces "Key Man" risk, which could make part of the enterprise dependent on an automated solution that is fragile, poorly documented, and lacks sufficient error handling, to name a few, in the worst examples of CD.

Having an Enterprise Development Mindset for Intelligent Automation is not just about implementing technology; it's about adopting a mindset that values stability, efficiency, and adaptability. The principles outlined in this chapter serve as a beacon, guiding organizations through the complex terrain of automation. In addition, by prioritizing good process understanding and discovery, organizations set themselves up for effective automations that align with their operational needs and Digital Transformation strategy.

The focus on automation reliance and stability, graceful error handling, and maintainability is crucial. These elements ensure that automated systems are efficient, reliable, and resilient in facing challenges and changes. This resilience is vital to maintaining operational continuity and achieving long-term success.

The path forward with IA principles is strategic growth and continuous improvement. It's about recognizing that automation is a tool and a transformative element that can propel organizations toward unprecedented efficiency, agility, and success when aligned with the right principles. As we embrace these principles, we pave the way for a future where automation is not just a part of the business process but a driving force behind its evolution and innovation. Real business cases are being devised that pair the incredible power of Traditional Machine Learning and, now, Generative AI with the astonishing versatility of RPA. Ensuring that Intelligent Automation development practices are sound, scalable, and sustainable has never had as much urgency as it does now.

Enhancing Intelligent Automation Through APIs and Database Integration

The reliance on GUI-based automation, adequate for certain tasks, unveils limitations when aspirations stretch towards scaling IA across an organization. The introduction of APIs into the automation ecosystem represents a critical step towards creating more robust, flexible, and scalable solutions. APIs enable seamless integration between disparate systems and applications, curtailing the dependency on surface-level, GUI-based interactions prone to inefficiency and breakage.

Similarly, the strategic leverage of databases in conjunction with IA initiatives augments data integrity and accessibility. Direct interactions with databases allow IA to efficiently manage data without the overhead of navigating through user interfaces. This not only increases processing times but also significantly reduces errors related to manual data entry or the higher exception rate associated with GUI-based automation, underscoring the transition towards more sophisticated automation paradigms.

Neglecting backend integration in IA initiatives leads to a plethora of challenges. Without direct database access and API usage, automation solutions may overly rely on GUI interactions. This reliance escalates the complexity and resource requirements for maintaining these automations, making the setup less scalable and error-prone, often resulting in higher computational resource consumption in terms of virtual machines, licenses, and other technical and non-technical terms. Consequently, this approach can lead to increased operational costs and diminished returns on investment in automation technologies.

Jeff Bezos' API 2002 mandate provides a strategic blueprint for organizations serious about Digital Transformation. Mandating all teams to expose their data and functionalities through service interfaces fosters an interconnected and collaborative environment. This ensures efficient team communication through shared services over the network, embodying the essence of a digitally transformed enterprise. The mandate's principles support a scalable, efficient digital ecosystem where integration and interoperability are paramount.

Adopting APIs and leveraging databases transcend mere technical decisions; they are strategic moves towards a digitally transformed business. This approach aligns with the overarching Digital Transformation goal: to reimagine business models, operating processes, and customer interactions for the digital age. As companies transition from basic RPA to sophisticated IA solutions that integrate APIs and databases, they lay the foundation for resilient, scalable, and efficient automation strategies that bolster critical business processes.

> **THE 2002 JEFF BEZOS MEMO**
>
> All teams will henceforth expose their data and functionality through service interfaces.
>
> Teams must communicate with each other through these interfaces.
>
> There will be no other form of interprocess communication allowed: no direct linking, no direct reads of another team's data store, no shared-memory model, no back-doors whatsoever. The only communication allowed is via service interface calls over the network.
>
> It doesn't matter what technology they use. HTTP, Corba, Pubsub, custom protocols — doesn't matter.
>
> All service interfaces, without exception, must be designed from the ground up to be externalizable. That is to say, the team must plan and design to be able to expose the interface to developers in the outside world. No exceptions.
>
> Anyone who doesn't do this will be fired.
>
> Thank you; have a nice day!

Figure 6.3, This is a reproduction of the memo Jeff Bezos famously wrote back in 2002 mandating that all systems within Amazon have native data integration capabilities. Many experts in Digital Transformation see it as one of the most monumental decisions he ever made and the one that enabled the Digitally Native Amazon we know today.

IT's Critical Role in Successful Intelligent Automation

A robust and adaptable Information Technology (IT) function is the engine of any successful Digital Transformation. As enterprises embrace strategies with Intelligent Automation at the core, it becomes even more crucial for application teams to step out of their traditional software development mindset and understand IA's deep dependency on their platforms.

One of the most fundamental distinctions that application teams must internalize is the vastly different nature of IA compared to conventional software solutions. Traditional software development often centers around building an application that serves a specific purpose. While a user might have flexibility and choices within the application, the software itself is designed to cater to a predefined scope. IA, on the other hand, focuses on automating existing tasks performed by humans. The key lies in understanding that a person completing a process often cuts across multiple applications, tools, and data sources.

Consider a simple example from the accounting world. An accountant might receive an invoice via email, manually extract data by eye, input it into an enterprise resource planning (ERP) application, and then file the invoice for archival purposes. Developing an IA solution here means automating these steps and bridging the gap between the email client, ERP system, and document management solution. The applications themselves don't change. The automation layer acts as the "glue," mimicking the actions the accountant took manually.

This inherent complexity demands a meticulous and coordinated approach from application teams to ensure IA endeavors are not derailed. A robust change control process is one of the most potent weapons in IT's arsenal to maintain stability and predictability. However, many legacy change control mechanisms are ill-equipped to address IA's realities fully. Intelligent Automation is acutely vulnerable to unexpected changes in the various applications they interact with. An update to the user interface of an ERP system, the rollout of a new version of a web portal, or even a seemingly minor patch to an email client can disrupt carefully built IA processes.

IT departments and application teams must proactively redefine their change control processes to treat IA as a first-class citizen. Any change, no matter how trivial it may appear, should be evaluated through the lens of its potential impact on existing automations. Notices of proposed changes have to inform stakeholders clearly about the potential implications for IA. Detailed information like updated field names, changes in button layouts, or even alterations in menu navigation flows are essential pieces of the puzzle for IA developers who need to safeguard the reliability of their automations.

Without this granular level of transparency and foresight, Digital Transformation efforts built upon IA will stumble repeatedly. Seemingly successful IA deployments suddenly malfunction due to an unforeseen change in an underlying application, becoming a source of widespread frustration for business users and eroding confidence in the entire Digital

Transformation push. Given this, the applications team's role becomes ensuring business continuity by acting as a shield for IA solutions, in the context of Digital Transformation.

The agility demanded by IA initiatives requires application teams to augment their change control with comprehensive regression testing and clear lines of communication between the application and IA teams. Often, IA developers are deeply familiar with the nuances of business processes but lack intimate knowledge of the intricate workings of all the applications their automations interact with. It is essential to establish a culture of collaboration where IT actively engages with IA teams to provide timely updates, technical insights, and support for proactive testing and adjustments well before changes go live in a production environment.

The success of a broad Digital Transformation strategy largely depends on how well IT organizations understand the core concepts of IA and create governance mechanisms that foster resilience. Leadership within the application teams, and the IT organization as a whole, need to become more than just knowledgeable about automation technology; they must internalize the implications and challenges of integrating IA into the fabric of the enterprise. It's about a mindset shift that embraces ongoing collaboration between application teams, IA experts, and business users for whom the automations are designed.

Let's illustrate this with a scenario from the world of insurance. An insurance company embarks on a project to automate a significant portion of its claims intake process. The IA solution interacts with the company's web portal, extracts data from uploaded documents, updates the core claims management system, and sends automated email notifications. Initially, things go smoothly; claim processing speeds up dramatically, customer satisfaction rises, and the investment seems to be paying off. However, without the application team's proactive involvement, a few months later, a significant upgrade to the claims management system introduced subtle layout changes to data entry fields. The impact is immediate, with the IA bot beginning to fail, causing delays, rework, and a negative impact on the very metrics that had shown such promise.

Now, contrast this scenario with one where the application team has effectively adapted for IA. Notices of upcoming changes to the claims management system would include explicit callouts for IA considerations. IA developers would be involved in testing the changes in a pre-production environment. They might even work with the application team to implement "hooks" or identifiers for crucial elements used by the automation, insulating them from future layouts or cosmetic alterations. Instead of disruption, the IA solution is seamlessly updated, avoiding any downtime.

Embracing IA means going beyond traditional development paradigms and technical proficiency. This means that application teams are taking a consultative role and partnering with business units to identify processes that are ripe for automation. They need to develop an in-depth understanding of the processes they seek to optimize with IA. This collaborative and process-centric approach becomes a catalyst within the broader Digital Transformation strategy.

The speed and responsiveness that the application teams bring to IA initiatives regarding application details are critical. Business agility in the current landscape often hinges on the adaptability of underlying processes. IA sits at the forefront of this revolution in process flexibility. Bottlenecks within the application team, cumbersome communication channels, and an inability to proactively address the needs of IA can significantly hold back the broader Digital Transformation goals of the company. On the other hand, when the application teams effectively champion IA, it becomes an exponential force that propels the business forward.

Informed leaders within the application teams understand that championing IA is not just having a program that automates some things here or there. It's about envisioning how technology can fundamentally alter how a business operates. They educate themselves on the potential and limitations of AI-infused automation. They build bridges between departments, fostering a culture where technology proactively enables business goals instead of reacting to demands. This strategic and forward-thinking approach within the application teams truly unleashes the transformative power of IA, allowing organizations to become faster, more responsive, and ultimately more competitive in terms of their Digital Transformation efforts.

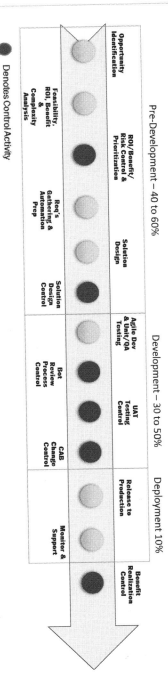

Figure 6.4, A typical RPA Bot Journey

Case Study: Embracing Enterprise Automation Principles

Two competing companies, TechnoEdge and OldTech Inc., embarked on their journey into the world of Intelligent Automation.

TechnoEdge, led by the visionary CIO, Emily Tran, embraced an Enterprise Automation mindset. Emily understood the importance of a thorough process understanding and discovery. She dedicated her team to meticulously mapping out the current process and to develop a robust solution and architectural design, before implementing any automation. This approach revealed many inefficiencies and bottlenecks that were previously invisible.

After learning all she could about the applications that she will be automating the process against, if available, Emily focused on backend integration and utilized APIs for deeper system connections, avoiding over-reliance on GUI-based automation. She implemented robust error-handling mechanisms, ensuring their automation solutions were efficient, modularly built, and resilient. This approach made their systems adaptable and easy to maintain, ready for the inevitable changes in technology and business processes.

The result was astounding. TechnoEdge's automations worked seamlessly, improving productivity and reducing operational costs. Emily's foresight in adopting Enterprise Automation principles set a standard in the industry, making TechnoEdge a beacon of innovation and efficiency. In addition, TechnoEdge redesignated its automation practice from an RPA CoE to an IA CoE to acknowledge its program's rapid growth and evolution over that period.

Contrastingly, OldTech Inc. took a more traditional approach under the leadership of its IT Director, John Smith. Skeptical about adding Enterprise Automation principles into his automations, John insisted on quick fixes using surface-level RPA solutions to get the quick win he felt Oldtech Inc. needed. His team hastily coded bots to automate tasks without deeply understanding the underlying processes or applications underpinning them.

OldTech's automations were primarily GUI-based, making them prone to errors and breakdowns when there were updates to the underlying applications. The bots were not equipped to handle exceptions gracefully, leading to frequent disruptions. Maintenance became a nightmare as his RPA team struggled to keep up with the constant need for updates and fixes.

The impact was immediate and severe. OldTech faced increasing operational costs, inefficiencies, and employee frustration. Their leadership and stakeholders started to notice the lack of reliability, leading to the enterprise's gradual loss of trust.

A year later, the contrast between the two companies was stark. TechnoEdge had successfully scaled up its operations, leveraging its stable and efficient automations. Emily's team continued to innovate, integrating ML-powered document understanding, Generative AI, and Traditional Machine Learning into their automation offerings, further enhancing their competitive edge.

OldTech's RPA CoE, however, found itself in a precarious position. Struggling to keep his team motivated and facing increasing skepticism, John met unhappy process owners and stakeholders and his senior leadership's growing suspicion of RPA's ability to deliver for OldTech. Realizing his mistake, he reached out to Emily for guidance. Recognizing the importance of Enterprise Automation principles, OldTech began its journey toward having an Enterprise Automation Development Mindset, albeit a little late.

Chapter 6 Key Points Recap: Fostering an Enterprise-Wide Automation Development Mindset for Intelligent Automation

- **Understanding IA's Fragility and the Need for Resilience:** IA solutions often face criticism for their fragility, especially in complex, mission-critical processes. An Enterprise Automation development mindset is crucial to enhance IA solutions' resilience, durability, and serviceability, ensuring they deliver value without the constant need for repairs and redeployment.

- **The Long-Term Perspective of IA Solutions:** Contrary to perceptions of IA as a temporary bridge solution, IA often represents a long-term solution capable of integrating data and completing tasks efficiently and reliably. This perspective demands IA solutions to be designed with long-term sustainability in mind.

- **Enterprise Automation Development Mindset:** This mindset involves viewing automation as an integral part of organizational technology, supported by a strategic vision that aligns with business goals. It emphasizes a holistic approach to automation, focusing on scalability, sustainability, and the optimization of end-to-end processes.

- **Importance of Process Understanding and Discovery:** A thorough understanding and discovery process is foundational for IA. It entails detailed process mapping, identification of process inefficiencies, and a deep dive into the applications involved, setting the stage for effective automation solutions.

- **Effective Solution and Architectural Design:** Strategic foresight in solution and architectural design is vital. Optimizing IA solutions for specific applications can significantly enhance automation efficiency, reduce integration complexity, and improve data accuracy.

- **Establishing Standards and Governance:** Developing standards based on reliability, stability, gracefulness, and maintainability and establishing governance mechanisms are essential for the long-lasting effectiveness of IA solutions. These standards ensure uniform quality and performance across all automation projects.

- **Embracing Citizen Development within a Structured Framework:** While citizen development democratizes automation, it must adhere to Enterprise Automation principles to avoid reinforcing negative perceptions of IA. Proper guidance and training can ensure that automations are built with sustainability in focus.

- **A Tale of Two Companies:** The contrasting journeys of TechnoEdge and OldTech Inc. illustrate the profound impact of embracing versus ignoring Enterprise Automation principles. TechnoEdge's success story highlights the benefits of a strategic, holistic approach to IA, while OldTech Inc.'s challenges underscore the pitfalls of neglecting such principles.

This chapter underlines the transformative potential of adopting an enterprise-wide automation development mindset in the realm of Intelligent Automation. It showcases how strategic, holistic approaches to IA can significantly contribute to the resilience and success of automation initiatives within the Digital Transformation journey.

In Chapter 7, We will discuss how IA can help an enterprise drive value and the importance of adopting or adjusting your operating model in order to achieve your Digital Transformation goals.

Chapter 7: How Intelligent Automation Can Be Key to Driving Strategic Value

How Strategies Using IA Can Help Move an Operating Model to Become Digitally Native

As discussed earlier in the Preface, the book by Tony Saldanha and Robert A. McDonald, "Why Digital Transformations Fail, The Surprising Disciplines of How to Take Off and Stay Ahead," the transformation of operating models to adapt to the emerging digital landscape is of the utmost strategic importance for businesses seeking to sustain and grow in an increasingly competitive environment. This is why the concept of being Digitally Native has been emphasized so far and will continue to be. A pillar of this transformation is a well-defined business strategy that outlines how a company intends to create and deliver value to its customers, setting itself apart from competitors. This strategy is intricately supported by an operating model that specifies the processes, technology, and human resources orchestrated to execute this strategic vision. The operating model acts as the mechanism through which a business's strategic objectives are operationalized, translating high-level goals into daily actions and interactions.

Intelligent Automation stands as a transformative force within this framework, offering a powerful toolset that extends beyond conventional automation to include Artificial Intelligence and NLP via IDP. This integration enhances process efficiency, fosters innovation, and elevates service delivery across the enterprise, promising to reshape business operations by:

- **Improving Customer Satisfaction:** With the pace of the world today, customer expectations are higher than ever. IA enables businesses to meet these expectations by automating customer service processes, ensuring quick responses are accurate and tailored to their needs. For example, chatbots powered by IA can handle a multitude of customer queries in real time, providing instant solutions or guiding customers through complex processes. This level of responsiveness and the ability to offer 24/7 service significantly enhance the customer experience, fostering loyalty and satisfaction. Additionally, by analyzing customer interaction data, IA can help businesses anticipate customer needs and tailor services or products, further enriching the customer experience.

- **Reducing Expenses and Increasing Revenues:** IA streamlines operations by automating routine and complex tasks, from data entry and processing to managing intricate supply chains. This reduces the need for manual intervention, which, in turn, lowers labor costs and minimizes the potential for costly errors. Furthermore, the efficiency gains from automation allow employees to focus on higher-value work, such as strategy and innovation, that can drive business growth which can boost revenue. Also, on the revenue side, IA opens up new business models and services opportunities. For instance, IA solutions powered by ML can unearth insights that lead to developing new products or enhancing existing offerings, tapping into previously unexplored markets or revenue streams.

- **Lower Operational Losses:** Manual processes are often riddled with inefficiencies and prone to errors, leading to rework, waste, and sometimes penalties or loss of business. IA mitigates these issues by ensuring processes are executed consistently and accurately. For example, in the Finance and Accounting departments, automating invoice processing or compliance reporting can significantly reduce the likelihood of errors that could result in financial losses or regulatory fines. By preemptively automating these processes, businesses can avoid the costs associated with correcting mistakes, not to mention the potential savings from avoiding litigation or compliance penalties.

- **Lowering Enterprise Risk:** The risk of human error is an ever-present concern in business operations, especially in areas requiring high precision or compliance with strict regulatory standards. IA minimizes this risk by automating critical processes, ensuring they are executed with consistent accuracy. Moreover, IA solutions can be configured to monitor for regulatory requirement changes, automatically alerting process owners when standards change so that the appropriate changes to their process can occur. This reduces the likelihood of compliance violations, regulatory fines, and other compliance issues. Additionally, IA can contribute to disaster recovery efforts by maintaining operational continuity through the automation of repetitive and standardized processes, further lowering enterprise risk in the face of unexpected events, as was seen during the COVID-19 pandemic mentioned earlier in the book.

Consider, for instance, the application of IA in the financial sector for automating processes like invoice processing, customer onboarding, and account management. The efficiencies gained not only enhance operational accuracy but also contribute to improved customer service. Similarly, IA's capability to provide real-time, personalized assistance in customer service scenarios underscores its potential to revolutionize engagement and satisfaction levels.

However, the essence of a Digital Transformation journey transcends the mere adoption of technological innovations. It necessitates a strategic alignment of business and operating models with digital capabilities. A business strategy that overlooks the integration of IA is fundamentally incomplete in the modern era. For example, relying on manual processes for tasks such as data reconciliation and reporting increases operational costs, slows down response times, and introduces a higher risk of inaccuracies. Incorporating IA into these processes can mitigate such risks, reduce operational costs, and enhance overall efficiency.

The revenue-generating capabilities of IA further underscore its strategic significance in Digital Transformation. IA can directly impact customer satisfaction and retention by enhancing customer engagement via automated service bots or streamlining order processing for faster delivery, leading to increased sales and revenue. Moreover, IA's ability to gather and process vast amounts of customer interaction data offers businesses critical insights for market strategy adjustment and new product development, directly linking operational efficiency with revenue growth opportunities.

The strategic integration and process automation powered by IA is perhaps its most compelling contribution to Digital Transformation. This transformation is about tactically gaining ROI and reimagining business operations for greater agility, innovation, and competitive edge. By embedding IA across processes and functions throughout the enterprise, businesses can achieve a systemic integration that aligns with and propels the company's overarching Digital Transformation goals. This strategic integration of IA within business operations fosters an environment of continuous improvement. It paves the way for incorporating more advanced AI capabilities in the future, ensuring the business remains at the forefront of technological advancement.

The role of IA within Digital Transformation strategies is both foundational and transformative. It offers a practical pathway to operational efficiency, cost management, and strategic business innovation. By adopting an integrated approach that positions IA at the heart of Digital Transformation efforts, businesses can ensure immediate improvements in efficiency and cost savings and lay the groundwork for sustained growth, innovation, and competitiveness.

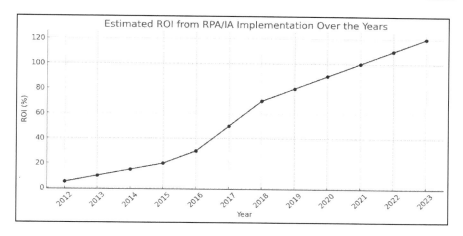

Figure 7.1, (OpenAI, n.d.) The graph above illustrates an estimated scenario for the Return on Investment (ROI) of RPA/IA technologies from 2012 to 2023 for a typical Fortune 2000 company following its Digital Transformation journey. This information was pieced together by Chat GPT-4 using publicly available information from companies like Everest Group, Gartner, or McKinsey based on what it calls "common findings and reported trends." Here's how the ROI progression can be interpreted:

- 2012-2013: These were the initial years with low ROI percentages (5-10%) as the company was just beginning to adopt RPA technologies and integrate them into its processes.
- 2014-2016: The ROI rose steadily (15-30%) as the company expanded its RPA use and saw significant cost savings and efficiency improvements.
- 2017-2020: Rapid growth in ROI (50-90%) marks the integration of AI with RPA, enhancing the capability to include Intelligent Automation and realizing higher benefits in both cost savings and strategic business outcomes.
- 2021-2023: The maturity phase shows ROI stabilizing at a high level (100-120%), indicating that automation is deeply embedded in the company's operational processes and contributing maximally to financial performance.

Using IA to Efficiently Bring Offshore Tasks Back Onshore Effectively

Integrating Intelligent Automation (IA) into organizational processes is a testament to the innovative strides companies are making toward Digital Transformation. One part of this movement is bringing offshore tasks back onshore efficiently, a trend accelerated by the unforeseen challenges discussed earlier in this book posed by the COVID-19 pandemic, highlighting a critical shift in operational strategy. With its closures and quarantines of offshore offices, the pandemic necessitated reevaluating how businesses maintain continuity and efficiency in the face of global disruptions. In this context, IA emerges as a solution and a strategic enabler of resilience and efficiency.

The economic argument for integrating IA into onshoring efforts is compelling. Software robots offer the ability to perform tasks at a fraction of the cost associated with offshore labor. The efficiency of these bots, characterized by their ability to operate with minimal error rates, directly translates into significant cost savings for businesses. By automating tasks traditionally outsourced offshore, companies can achieve a dual objective: maintaining operational efficiency and reducing expenditure. Furthermore, the lower error rates inherent in IA solutions address a critical aspect of business operations - risk mitigation.

Mitigating risks, especially those related to compliance and regulatory standards, becomes a more manageable endeavor with IA. The precision and reliability of automated processes ensure adherence to regulatory requirements, minimizing the potential for mistakes that could lead to financial repercussions, legal trouble, or both. This aspect of IA is particularly crucial in industries where regulatory compliance is non-negotiable. Moreover, the reputational benefits of deploying IA, underscored by enhanced reliability and accuracy, cannot be overstated. In a market where customer trust is paramount, consistently delivering accurate and compliant services is a competitive advantage.

Client satisfaction, a keystone of business success, is profoundly influenced by the speed, accuracy, and reliability of service delivery. The agility afforded by IA allows companies to rapidly adapt processes in response to customer feedback, providing a dynamic and responsive service delivery model. This emphasis on customer satisfaction positions the company as a responsive and customer-focused organization, which fosters customer loyalty.

Beyond the tactical benefits of cost reduction and risk mitigation, the strategic significance of incorporating IA into onshoring efforts lies in its role as an enabler of Digital Transformation. The emphasis on automated solutions, characterized by complex processes, represents a fundamental shift in how businesses view and leverage technology. By adopting IA, companies are not merely responding to immediate operational challenges, but are laying the groundwork for a future where Intelligent Automation can become a crucial driver of business strategy, innovation, and growth. Additionally, the agility offered by IA, where processes can be swiftly adapted to meet changing regulatory realities and customer expectations, positions organizations to navigate the complexities of the modern business environment with unprecedented flexibility.

Examples of Onshoring Work from Offshore Using IA

Let's journey through some hypothetical scenarios across diverse industries using IA to onshore work within specific industries. While not extracted from specific real-world cases, these illustrative examples vividly show how IA can revolutionize business operations, streamline processes, and contribute to strategically onshoring previously outsourced tasks and roles.

Imagine a scenario within the financial services sector where a leading bank harnesses IA to transform its customer service framework. Traditionally dependent on offshore resources to manage inquiries, this bank integrates IA to automate responses to common questions. This strategic application elevates customer service efficiency and enables the repatriation of certain roles, fostering job creation domestically to handle more complex issues by human beings where cultural familiarity is a key differentiator. As the IA solution continuously learns and evolves, it exemplifies the potential of automation in enhancing service quality and adaptability in customer interactions.

Venturing into the healthcare domain, we envision a healthcare provider leveraging IA to refine patient data management. Once managed by an extensive offshore team, automating record processing and updates underscores the dual advantages of operational streamlining and enhanced data security. By minimizing human intervention in sensitive data handling, this hypothetical example showcases IA's capacity to bolster data accuracy, privacy and vulnerability, illustrating the transformative impact of IA on healthcare operations.

Consider a fictional scenario in the insurance industry where an insurance company adopts IA for claims processing. Automating the initial stages, such as data entry and document verification, accelerates the process and allows for the onshoring of more nuanced decision-making roles. This strategic move enhances efficiency and deepens customer trust and satisfaction, illustrating IA's potential to reallocate human resources towards more impactful, strategic tasks.

While conceptual, these scenarios underscore IA's transformative potential across various sectors. They highlight the strategic use of automation to improve operational efficiencies and enable the strategic onshoring of tasks, contributing to innovation and job creation. At the same time, the journey towards integrating IA into Digital Transformation strategies encompasses challenges, including the need for robust governance and continuous adaptation. The envisioned benefits across these hypothetical examples illuminate IA's role in driving forward-thinking business outcomes. As we continue to explore the possibilities of IA, these illustrative examples serve as a testament to the technology's transformative power, emphasizing the importance of strategic integration and alignment with overarching business objectives.

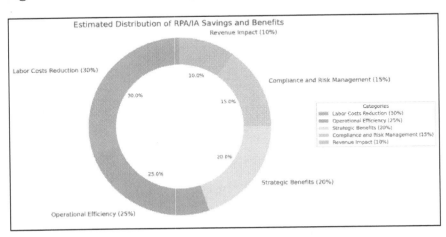

Figure 7.2, (OpenAI, n.d.) The pie chart above illustrates the estimated distribution of savings and benefits from RPA (Robotic Process Automation) and IA (Intelligent Automation) within a typical organization. Here's the breakdown for each category:

Breakdown of Savings and Benefits:

- Labor Costs Reduction (30%)
    - Human Resources (15%): Savings from automating tasks such as payroll processing, employee data management, and recruitment activities.
    - Operations (15%): Savings in operational departments like Finance and Accounting through the automation of transactional tasks such as invoicing and claims processing.
- Operational Efficiency (25%)
    - Process Improvement (15%): Efficiency gains from process standardization and optimization, which improve the throughput and quality of outputs.
    - Resource Utilization (10%): Better utilization of assets and infrastructure, reducing overhead costs and enhancing operational capacity without proportional increases in expenditures.
- Strategic Benefits (20%)
    - Business Agility (10%): Enhanced flexibility and innovation capability, allowing the company to adapt to changes and seize new opportunities quickly.
    - Customer Satisfaction (10%): Improve service levels and customer experience through faster response times and higher accuracy, enhancing overall customer satisfaction and loyalty.
- Compliance and Risk Management (15%)
    - Regulatory Compliance (10%): Automation helps ensure consistent execution of compliant processes and more accessible auditability, reducing non-compliance risks and associated penalties.
    - Risk Reduction (5%): Decreased the likelihood of data breaches and errors through automated data handling and improved security protocols.
- Revenue Impact (10%)

- Market Opportunities (5%): Faster time to market for new products and services and enhanced capabilities for cross-selling and up-selling.
- Improved Decision Making (5%): Leveraging AI-enhanced analytics to provide deeper insights for better strategic decisions.

This visualization helps stakeholders understand where to focus their efforts for maximum impact when implementing RPA and IA technologies. It reflects typical outcomes in enhanced efficiency, cost savings, and strategic business performance. However, please keep in mind that these percentages could be vastly different for your company depending on its culture, current level of digital maturity, and industry.

Case Study: Crimson Air's IA Journey

As he adjusted the microphone, the fluorescent lights glared in Jonathan Jet's eyes. A hush fell over the room filled with analysts and reporters. Two and a half years ago, the merger that formed Crimson Air had been hailed as a turning point for the airline industry. Yet, the path since then had been anything but smooth.

"Thank you for joining our company's quarterly earnings call today," Jonathan began, his voice steady. "Eighteen months ago, I stood here and shared the challenges we faced. Despite promises of streamlined operations and increased efficiencies, our costs remained stubbornly high. Much of this stemmed from legacy systems, our siloed operations for core functions, a reliance on traditional AI that couldn't handle our growing complexity, and a significant amount of work offshored to reduce costs."

He paused, recalling the tense board meetings, the anxious reports, the mounting pressure.

"But we were determined to fly higher," he continued. "We embarked on a bold Digital Transformation strategy, placing Intelligent Automation – specifically a combination of RPA, updated AI, and burgeoning Generative AI capabilities – at its very core. This wasn't about mere offshore cost-cutting; it was about reimagining the very way we operate and bringing value back to our domestic operations."

A slide flickered onto the screen behind him, detailing Crimson Air's strategic pillars.

"Our goal was to become a truly Digitally Native airline," Jonathan explained. "That meant leveraging technology to drive operational excellence, enhance customer experience, unlock new revenue streams, and onshore work wherever it made strategic sense."

He pointed to the screen. "Intelligent Automation was the linchpin. RPA streamlined existing operations both on and offshore. Bots powered by traditional AI, updated with better training data from the merger, optimized processes across the board. But it was Generative AI that supercharged our onshoring transition. Models trained on both our offshore operations and domestic goals could translate the nuances of work – customer conversations, ticketing flows, and even back-office tasks – allowing us to seamlessly transition that knowledge to our new domestic staff. This wasn't about replacing staff; it was about empowering them."

"Mr. Jet," an analyst from SilverOak Capital piped up, a skeptical edge to his voice, "these are lofty goals. But how did this translate into tangible financial results?"

Jonathan smiled. He'd been expecting this. "The numbers speak for themselves, Mr. Thompson. In the past twelve months, we've seen a significant improvement in operating leverage. The IA-powered onshoring transition, while initially an investment, has already delivered long-term savings. Our expense ratio has decreased by 15%, and customer satisfaction metrics have gone up due to smoother communication and faster domestic response times."

An approving murmur swept through the room.

"But it's not just about savings," Jonathan added. "Generative AI isn't just about automation, it's about insight. We use it to analyze vast datasets, identify new market trends, and develop offers that feel tailor-made for our customers. This intelligent personalization has driven revenue growth in key segments."

Another analyst, a woman from Blackwood Investments, raised her hand. "Mr. Jet, while these results are impressive, there's always the concern of technological disruption in a fast-paced industry like aviation. How do you ensure Crimson Air stays ahead of the curve?"

"A valid question, Ms. Carter," Jonathan acknowledged. "Our Digital Transformation journey is continuous. We've established a dedicated R&D team that explores cutting-edge technologies like predictive analytics, virtual reality for training, and even Blockchain for supply chain management. We partner with universities and AI labs, ensuring we're always on the edge of what's possible. This, combined with our empowered workforce, both onshored and in strategic offshore locations, means we're not just following trends but shaping them."

The questions continued at this pace throughout the call. Jonathan discussed the challenges, the measured rollout of their onshoring program, and Intelligent Automation with its blend of RPA and AI, along with clear success metrics.

As the call concluded, a sense of renewed optimism was in the air. Crimson Air had weathered the turbulence, and the journey ahead looked promising. The merger, once faltering, had finally taken flight, propelled by the power of Digital Transformation and a strategic blend of domestic operations and targeted offshoring.

Chapter 7 Key Points Recap: Leveraging Intelligent Automation for Strategic Transformation

- **Strategic Importance of Intelligent Automation:** IA combines advanced technologies such as AI and NLP to enhance operational efficiency and drive innovation. This positions IA as a crucial component of Digital Transformation strategies aimed at reshaping business models for the digital age.
- **Enhancing Customer Experience:** IA improves customer satisfaction and loyalty by automating and personalizing customer interactions. Tools like AI-driven chatbots provide real-time, 24/7 customer support, vital in today's fast-paced market environment.
- **Financial Optimization:** IA reduces costs by automating routine and complex tasks, which decreases the need for manual labor and minimizes errors that can lead to financial losses. Additionally, it unlocks new revenue streams through innovative business models and product enhancements informed by Machine Learning analytics.
- **Risk Reduction and Compliance:** Automation through IA ensures high accuracy in processes requiring regulatory standards compliance, reducing the risk of fines and penalties. It also supports disaster recovery

plans by maintaining business continuity through automated systems.

- **Operational Efficiency and Strategic Onshoring:** IA streamlines operations, allowing businesses to adapt processes quickly and efficiently, which is particularly beneficial in reducing reliance on offshore operations and bringing tasks back onshore, thereby enhancing control and responsiveness.
- **Supporting Strategic Business Goals:** The deployment of IA is integral to refining business strategies that focus on innovation, agility, and customer-centricity. It aligns digital capabilities with business objectives, ensuring companies stay competitive and respond to market dynamics.
- **Case Study: Crimson Air:** Crimson Air's story illustrates the real-world application and benefits of IA post-merger. The airline's successful IA deployment showcases improved operational leverage, customer satisfaction, and strategic repatriation of tasks, affirming the critical role of IA in achieving a Digitally Native operating model.

In essence, Chapter 7 positions Intelligent Automation as a cornerstone of strategic business transformation, driving operational excellence, reducing risks, and laying the groundwork for a digital native future.

As we transition to Part 3 of the book, we will discuss the importance of Intelligent Automation in crafting a Digital Transformation strategy based on principles that align with a company's vision, values, and mission.

Part 3: The Importance of Intelligent Automation in Crafting a Digital Transformation Strategy

Chapter 8: Aligning IA with Digital Transformation Principles

The Importance of Vision, Values, and Mission in Crafting any Digital Transformation Strategy

Aligning with the firm's vision, values, and mission in crafting any Digital Transformation strategy is fundamental to ensuring that the initiative resonates with the organization's core purpose, guiding it toward a future where it remains relevant. This alignment is critical for operational efficiency or market competitiveness and for embodying the organization's identity and aspirations.

At the center of creating a successful Digital Transformation strategy is the understanding that this journey is more than the adoption of new technologies; it's a comprehensive reimagining of how an organization operates, delivers value, and engages with its stakeholders. For example, a firm committed to sustainability can integrate this value into its Digital Transformation by prioritizing green technologies and digital processes that minimize environmental impact, reflecting its core values in its operational practices and offerings. In another example, a financial institution committed to fair and equitable lending ensures that the data used for its Machine Learning models is not only not biased based on race or ethnicity but also has models specifically designed to optimize products for historically disadvantaged groups.

Determining the organization's aspiration to become Digitally Native requires thoroughly reevaluating the current business strategy and operating model. This involves understanding the necessary changes for the organization to survive and thrive. Intelligent Automation is crucial in this transition, offering insights and capabilities that redefine the operating model. IA can help streamline operations, enhance customer engagement through personalized digital experiences, and foster innovation. Additionally, the strategic integration of emerging technologies like Cloud Computing and Blockchain into the operating model presents new efficiency, security, and innovation opportunities.

Transitioning to a Digitally Native stance necessitates a holistic approach encompassing people, processes, governance, and technologies. This includes rethinking workforce skills and culture to embrace digital literacy, reengineering processes to be agile and customer-focused, establishing robust governance structures to steer digital initiatives effectively, and leveraging the right technologies to drive transformation forward. An effective change management program is crucial in this journey, as it addresses resistance, nurtures a culture of learning and innovation, and ensures that Digital Transformation initiatives are aligned with strategic business goals.

For instance, consider a service firm automating its customer service processes using AI and Machine Learning. This move, aimed at improving efficiency, is also designed to deliver personalized service experiences, aligning with its mission of exceptional customer care. Such initiatives exemplify how Digital Transformation, rooted in the organization's vision, values, and mission, can significantly enhance operational capabilities and customer engagement.

By embedding IA and other digital technologies into the organization's fabric, firms can streamline operations and unlock new innovation and value-creation avenues that resonate with their foundational principles. This holistic and aligned approach to Digital Transformation fundamentally transforms an organization, making it more resilient, agile, and innovative in the face of digital disruption.

Principles of Digital Transformation and the Role of Intelligent Automation

The successful integration of Digital Transformation strategy principles is crucial for companies looking to innovate and stay competitive. These principles offer a blueprint for digital integration and ensure that such efforts are strategically aligned and sustainably implemented within the organization.

Enhancing the Customer Experience: This is one of the key principles and leverages digital solutions like AI-powered services and upgraded online platforms. This principle is about making interactions more intuitive and responsive, which can be directly supported by Intelligent Automation (IA) through technologies like chatbots and personalized online services that operate efficiently and reduce response times.

Enabling Data-Driven Decision-Making: This is another critical principle, focusing on leveraging big data and Advanced Analytics to drive business strategies. IA supports this principle by automating data collection and processing, thus providing real-time analytics that help uncover market trends and customer preferences.

Cybersecurity and Compliance: These principles are more relevant than ever as Digital Transformation expands its digital footprint across your company and across industries. IA contributes to this principle by automating compliance checks and maintaining security protocols that protect sensitive data and ensure that operations adhere to regulatory standards without human error.

Fostering a Culture of Innovation: This is facilitated by IA by automating routine tasks and freeing up human resources to engage in more innovative activities. This principle encourages a mindset shift within the organization, promoting creativity and experimentation, which are essential for innovation.

Continuous Improvement: This principle is dedicated to enhancing business processes and strategies that benefit from IA's ability to provide feedback and insights quickly. Organizations can use IA to continually refine and optimize their digital strategy and operations based on accurate and up-to-date data.

Establishment of a Dedicated Transformation Team: This team is central to driving the Digital Transformation efforts across the company, ensuring that every initiative aligns with strategic business goals. As discussed earlier in the book, the role of a Digital Transformation Office (DTO) could be instrumental here. This team works closely with technology and business leaders to integrate digital technologies such as IA into everyday business processes, ensuring that these technologies are implemented and effectively contributing to the company's digital maturity.

Integrating Intelligent Automation within these principles enhances their implementation by providing scalable solutions that support current operational needs and future demands. IA's role in automating processes, enhancing data analysis, and supporting continuous improvement initiatives makes it a critical factor in the successful adoption of Digital Transformation strategies. This integration ensures that businesses keep up with digital trends and lead in innovation, efficiency, and customer satisfaction.

Ten Digital Transformation Strategy Principles

Enhance the Customer Experience
- Implement digital solutions like upgraded online platforms and AI-powered services to improve customer interaction and service

Enable Data-Driven Decision Making
- Data will be central to decision-making processes and will be critical to unveiling market trends and customer preferences

Foster a Culture of Innovation
- Promote a company-wide culture of innovation through internal campaigns, workshops, and incentives for creative ideas

Enhance Cybersecurity and Compliance
- Protect customer's data and compliance with regulatory standards by investing a robust cybersecurity framework and ensuring our efforts align with regulatory requirements

Focus on Continuous Improvement
- Encourage a continuous improvement mindset, actively seeking and using feedback to refine digital strategies and operations

Increase Operational Efficiency
- Streamline operations by automating routine tasks to digitizing traditional processes while enhancing efficiency and reducing operational costs

Increase Employee Training and Development
- Launch training programs for employees in digital skills, including software use, data analysis, and cybersecurity awareness, and increase digital literacy throughout the company as a whole

Establish a Dedicated Transformation Team
- Form a cross-functional team with experts in digital tech, data analytics, customer experience, cybersecurity, and change management to drive digital transformation efforts throughout the company

Engage Partnerships and Collaborations
- Form strategic partnerships with tech companies, fintech startups, and other financial institutions for access to new technologies and business models

Work Towards Being Digitally Native
- Commit to digital initiatives that align with long-term goals and strategy to ensure that digital transformation contributes to enduring success

Figure 8.1, (Saldanha, 2019), A set of Enterprise Digital Transformation Principles

The Importance of Having the Right Organizational Incentives for Effective Digital Transformation

A critical factor for success is ensuring the right organizational incentives are in place from top to bottom. Digital Transformation is a complex and multifaceted endeavor that requires the entire organization to be aligned and committed to the process. Without the proper incentives in place, the chances of ever becoming Digitally Native are significantly diminished.

The importance of having the right organizational incentives starts at the very top of the organization. The vision and strategic direction for Digital Transformation must come from the top leadership, and this vision must be clearly communicated and cascaded throughout the entire organization. This top-down accountability is crucial in developing a compelling vision statement for the Digital Transformation journey. The company's vision statement should align with the existing corporate mission and business objectives and serve as a guiding light for the entire organization.

For example, consider the case of a large financial services firm embarking on a Digital Transformation journey. The company's leadership recognized the need to modernize its operations and enhance the customer experience by adopting new technologies and digital capabilities. The initial step was to develop a clear vision statement aligned with the company's corporate mission and strategic priorities.

The vision statement emphasized the importance of becoming a truly Digitally Native organization, focusing on delivering seamless, personalized experiences to customers through Advanced Analytics, Artificial Intelligence, and other cutting-edge technologies, along with optimizing the company's internal operations using Intelligent Automation via RPA, IDP, and both Traditional and Generative AI. This vision statement was then cascaded throughout the organization, with the C-Suite taking a lead role in communicating the strategic direction and securing the necessary buy-in from the company's various business units and functional teams.

Developing a formal Digital Transformation strategy is a critical next step, and this strategy must be crafted with the vision statement in mind. The strategy should outline the specific goals, objectives, and initiatives that will be undertaken to achieve the desired Digital Transformation outcomes. Importantly, the company's C-Suite must drive and execute this strategy, with the Board of Directors providing the necessary oversight and accountability.

By aligning the organization's incentive plans with the Digital Transformation strategy, the company can foster a culture of innovation and continuous improvement, where employees are motivated to explore new technologies, experiment with new ways of working, and ultimately drive the organization toward its Digital Transformation goals.

However, it is essential to note that the incentive plans must be carefully designed to ensure they do not inadvertently encourage the wrong behaviors, such as focusing solely on cost-cutting measures that could undermine the company's long-term competitiveness. As discussed before, the incentive plans must be "smart" and sustainable, emphasizing driving innovation, enhancing customer experiences, and building a Digitally Native organization.

Alongside the top-down accountability and vision-setting, it is also crucial to ensure that "Bottom-Up Responsibility" is in place. This involves empowering the firm's lines of business and business functions to develop their own visions for product innovation and operational process improvements within their respective domains. By fostering this Bottom-Up Responsibility, the organization can harness the deep domain expertise and insights of its frontline teams, ensuring that the Digital Transformation efforts are grounded in the realities of the business and responsive to the evolving needs of customers and stakeholders.

For example, a healthcare provider embarking on a Digital Transformation journey could develop a vision statement that emphasizes the importance of using digital technologies to improve and enhance patient outcomes and experience, and drive greater operational effectiveness across the organization.

As the Digital Transformation strategy is formalized, the leadership team begins to recognize the need to empower the various business units and functional teams to contribute to the transformation efforts. The hospital's outpatient clinic teams could be encouraged to explore ways to streamline the patient check-in process and utilize data analytics to proactively identify and address patient needs.

Similarly, the organization's back-office teams, such as Finance and HR, could identify opportunities to digitize and automate their respective processes, reducing manual tasks and enhancing the organization's overall efficiency. By fostering this Bottom-Up Responsibility, the healthcare provider can tap into its workforce's collective intelligence and expertise, ensuring that the Digital Transformation efforts are specified by the needs and challenges faced by each business unit and function.

Notably, the incentive plans for these teams were also aligned with the overall Digital Transformation strategy, focusing on metrics that would drive the desired behaviors and outcomes. For example, the outpatient clinic teams were incentivized based on their ability to increase patient satisfaction scores, reduce wait times, and expand the adoption of digital health tools among their patient population. On the other hand, the back-office teams were evaluated based on their ability to streamline processes, reduce hours spent on manual tasks, and enhance the overall efficiency of their respective functions.

By aligning the organizational incentives from top to bottom, the healthcare provider will be able to create a culture of shared ownership and accountability for the Digital Transformation journey. This, in turn, ensures that the transformation efforts are not simply viewed as an IT-driven initiative but a holistic, organization-wide endeavor that requires the active participation and engagement of all employees.

As the organization progresses with its Digital Transformation, ensuring that the incentive plans and compensation structures were designed to encourage the desired behaviors and make the company more resilient and resistant to disruption became increasingly important. This meant that the incentive plans needed to go beyond just short-term cost savings or revenue growth, and instead focus on building long-term competitive advantages through developing new digital capabilities, enhancing customer experiences, and streamlining operational processes.

The incentive plans for the organization's product development teams could be designed to reward the successful launch of new digital products and services and the degree to which these offerings increased customer loyalty, expanded the company's addressable market, and strengthened its position against potential competitors. Similarly, the compensation for the organization's IT and operations teams can be tied to metrics that measure the impact of their Digital Transformation efforts on the company's overall resilience and ability to adapt to changing market conditions.

By aligning the incentive plans to build a truly Digitally Native and disruption-resistant organization, the company can foster a culture of strategic thinking and long-term value creation, rather than simply chasing short-term gains. This, in turn, helps to ensure that the Digital Transformation efforts are sustainable and position the company for continued success in the years to come.

Ultimately, the success of any Digital Transformation effort hinges on the organization's ability to align its incentive structures with the strategic vision and objectives. This necessitates a careful balance, ensuring that the incentive plans do not inadvertently encourage the wrong behaviors or undermine the organization's long-term competitiveness.

The importance of having the right organizational incentive plans at your organization from top to bottom in place cannot be overstated. By fostering a culture of accountability, empowerment, and shared ownership, organizations can increase the likelihood of their Digital Transformation initiatives achieving the desired outcomes and positioning the company for sustained success.

Metrics for Guiding Successful Digital Transformation

Effective Digital Transformation is fundamentally about reimagining how businesses operate and deliver value to customers. It is not just about technology, but also about aligning and incentivizing the entire organization towards this goal. Without the right metrics, organizations risk incentivizing short-term gains that may look good on paper but potentially harm the company's long-term health and competitive edge.

To ensure that the company's Digital Transformation initiatives align with long-term strategic objectives, businesses must develop a balanced set of metrics. These metrics should measure immediate operational efficiency and gauge long-term innovation and market adaptability.

Here are a few metrics that you might want to consider for this purpose:

- **Net Promoter Score (NPS):** This metric evaluates customer satisfaction and loyalty by measuring the likelihood of customers recommending a company's product or service to others. A high NPS indicates that the company's Digital Transformations are improving customer experiences in meaningful ways.
- **Customer Retention Rate:** This metric helps businesses understand the effectiveness of their

customer engagement strategies, a critical aspect of Digital Transformation. Retention rates can indicate the success of digital tools and customer relationship management systems in maintaining customer loyalty.

- **Time to Market:** The speed at which new products or services are developed and launched can be a critical indicator of a successful Digital Transformation. This metric reflects the organization's agility and ability to innovate.
- **Employee Productivity Metrics**: By tracking the output per employee before and after digital initiatives, companies can assess the direct impact of these transformations on operational efficiency (can employees effectively do more post-digital implementation).
- **Automation Efficiency Metrics:** These could include measures like the automation rate, which reflects the percentage of processes automated (via RPA, IDP, ML, or a combination of IA capabilities into one digital solution) and the impact of these automations on reducing cost and time while increasing efficiency and effectiveness.
- **Innovation Impact Score:** Measures the impact of new digital initiatives on the market and the organization, assessing factors like new revenue streams created and new markets reached.

- **Transformation-Specific ROI:** For every major digital initiative, calculating the return on investment can provide clarity on the financial benefits of the transformation efforts relative to the costs incurred.

Using these metrics, organizations can foster a culture that values both efficiency and innovation, discouraging decisions that sacrifice long-term growth for short-term efficiency gains. Metrics like NPS, customer retention, and innovation impact score ensure that transformations enhance competitiveness and customer satisfaction, rather than just cutting costs.

The strategic use of comprehensive metrics is crucial in guiding and evaluating the success of Digital Transformation efforts. They help ensure that the organization remains aligned with its long-term goals while adapting to digital advancements, thus supporting sustainable growth and competitive advantage.

Case Study: Venerable Bank Battles to Reinvent Itself in the Digital Age

Old Money Financial, a long-standing institution in the heart of Chicago, found itself at a critical crossroads. For decades, the bank had been a pillar of the Midwestern financial landscape, built on a foundation of tradition, stability, and conservative values. However, as the AI-led revolution swept through the industry, Old Money found itself falling behind the rapid pace of change, struggling to keep up with the agile FinTech startups and the Digital Transformation efforts of its more modern competitors.

In this context, the bank's Board of Directors gathered for a series of pivotal meetings to chart a new course for Old Money Financial. The boardroom was a reflection of the industry's shifting landscape, with members hailing from diverse backgrounds – some seasoned bankers steeped in the old ways, others tech-savvy disruptors eager to challenge the status quo.

At the forefront of this group was Charles Erikson, the bank's newly appointed CEO, a seasoned veteran with a reputation for bold moves. Recognizing the urgency of the situation, he had assembled a team of directors with a wide range of perspectives, hoping to forge a comprehensive Digital Transformation strategy that would propel Old Money into the future.

Seated at the table was Emily Greenfield, a former executive at a leading FinTech startup. Her background in innovative technologies and disruptive business models brought a fresh, forward-thinking approach to the discussions. Across from her, John Simmons, a retired banker with over four decades of experience, eyed her skeptically, his brow furrowed with concern.

"This Digital Transformation and AI nonsense is just a fad," Simmons huffed, his voice carrying an air of dismissal. "We've weathered storms before, and we'll do it again the old-fashioned way with determination and grit. Technology isn't the answer – it's our people and our values that have made this bank great."

Emily's eyes narrowed as she leaned forward, her expression steely. "With all due respect, Mr. Simmons, the world is changing, and if we don't adapt, we'll be left behind. FinTech companies are eating our lunch, and our traditional competitors are outpacing us and making strides that we simply can't ignore."

The tension in the room was palpable as the two clashed, their contrasting perspectives reflecting the very heart of Old Money's dilemma. CEO Erikson, sensing the need to find common ground, interjected.

"Both of you make valid points," he said, his voice calm and measured. "The truth is, we need to find a way to balance our core values and heritage with the demands of the modern financial landscape. This isn't an either-or situation – it's about finding the right blend of tradition and innovation."

As the discussion progressed, the board members grappled with the complexities of reinventing Old Money's business and operating model. They debated the merits of embracing new technologies, streamlining processes, and cultivating a more agile, collaborative culture.

Representing the middle ground was Sarah Watkins, a seasoned executive who had overseen the Digital Transformation of a large manufacturing conglomerate. Her insights on the people-centric aspects of change management proved invaluable as the board wrestled with the human element of their digital journey.

"It's not just about the technology," Watkins emphasized. "We need to invest in our people, equipping them with the skills and mindset to thrive in a digital-first environment. And we have to be willing to challenge our hierarchical structures and embrace a more collaborative, adaptive way of working."

As the meetings progressed, the board members began to see the value in each other's perspectives. The retired bankers, while initially skeptical, recognized the urgency of embracing Digital Transformation to remain competitive. The tech-savvy disruptors, in turn, gained a deeper appreciation for the bank's rich history and the importance of preserving its core identity.

Gradually, a shared vision began to emerge – one that combined the best of Old Money's traditional strengths with the agility and innovation required to succeed in the digital age. The board members, once divided, now found themselves united in their commitment to reimagine the bank's future.

Yet, as the final meeting drew to a close, a sense of uncertainty lingered. The path forward was still fraught with challenges, and the board members knew that the real work had only just begun. Old Money Financial's Digital Transformation would require unwavering dedication, a willingness to embrace change, and a steadfast commitment to navigating the uncharted waters of the modern financial landscape.

With a collective deep breath, the directors departed, their minds racing with the weight of the decisions that lay ahead. The future of Old Money Financial hung in the balance.

Chapter 8 Key Points Recap: Aligning IA with Digital Transformation Principles

- **Strategic Alignment with Core Values:** Effective Digital Transformation strategies are deeply aligned with an organization's vision, values, and mission, ensuring that digital initiatives propel the company towards future relevance and success.
- **Beyond Technology A Comprehensive Reimagining:** Digital Transformation transcends technology adoption, requiring a comprehensive reevaluation of the business strategy, operational models, and stakeholder engagement, emphasizing sustainability and inclusivity.
- **Holistic Approach and Adaptive Change Management:** Achieving digital maturity demands a holistic approach that includes redefining workforce

capabilities, enhancing process agility, and establishing robust governance supported by change management programs that foster a culture of continuous innovation.

- **Incorporating Intelligent Automation:** Integrating Intelligent Automation into the digital framework from the outset marks a shift from traditional processes to streamlined, automated operations, emphasizing the role of enterprise-wide automation in sustainable business strategy.

- **Incentivizing Organizational Change:** The success of Digital Transformation efforts hinges on having the right organizational incentives aligned with digital goals, promoting a culture of shared accountability, empowerment, and long-term strategic vision.

- **Importance of Strategic Metrics**: Effective Digital Transformation involves more than technology; it requires aligning business operations and enhancing customer value. Strategic metrics are essential for ensuring these initiatives align with long-term goals and support market adaptability.

- **Old Money Financial's Transformation Journey:** Old Money Financial's story illustrates the challenge of balancing tradition with innovation, highlighting the critical need for established institutions to embrace Digital Transformation principles to navigate the complexities of the modern financial landscape effectively.

By embedding IA and other digital technologies into the organizational framework and aligning these efforts with the company's foundational principles, firms can unlock new avenues for innovation and value creation.

In Chapter 9, we will discuss how IA can help firms be better positioned to handle digital disruptions.

Chapter 9: Sustaining Being Digitally Native Amid Rapid Tech Disruptions

The Essential Role of Being Digitally Native in Modern Business

As Digital Transformation advances, being Digitally Native is not just an advantage; it's a necessity for survival and growth. This term, mentioned in prior chapters in the book, emphasizes the innate ability of organizations to seamlessly integrate digital technologies into their core business processes and strategies. It's a mindset that enables firms to thrive in a world where technological disruptions are not exceptions but constants.

Over the last two to three years, the pace of innovation has accelerated dramatically, introducing groundbreaking developments across various sectors. For example, Intelligent Automation has been leveraged in the financial services industry to revolutionize customer service through ML-driven chatbots that respond to customer queries in real-time and predict and initiate conversations based on customer behavior and preferences. These AI-driven assistants can process and analyze vast amounts of data, offering personalized banking advice, much like a human advisor would, but with greater focus and availability.

In the healthcare sector, Intelligent Automation has facilitated the seamless integration of patient records across platforms, enabling healthcare providers to offer more coordinated care. A possible use case could involve an automated system that tracks patient health data from wearable devices, updates electronic health records in real-time, and alerts medical professionals to potential health issues before they become serious, thereby improving patient outcomes and reducing healthcare costs.

However, the rapid evolution of technology also presents significant challenges. The landscape is littered with firms that fail to adapt quickly enough to these changes, suffering a decline in competitiveness or even ceasing operations. A notable example is a company in the retail industry that neglected the shift towards e-commerce and digital marketing. Despite having a strong brick-and-mortar presence, the company struggled to compete with Digitally Native competitors that offered online shopping experiences, personalized customer engagement through data analytics, and efficient home delivery services. The failure to adopt a digital-first strategy led to dwindling sales, store closures, and, eventually, bankruptcy.

Conversely, firms that have embraced Digital Transformation are reaping substantial benefits. A company in the legal industry developed an Intelligent Automation solution for document understanding, drastically reducing the time lawyers spend reviewing legal documents. The system uses Natural Language Processing and Machine Learning to identify relevant information and assess legal arguments based on the understood legal precedence. This improves efficiency and accuracy and allows lawyers to focus more on legal analysis and strategy, enhancing the value they provide to their clients (Ajay Agrawal J. G., 2022).

These examples underscore the essence of being Digitally Native and having the ability to continuously innovate and adapt in response to technological advancements. Intelligent Automation plays a crucial role in this context.

However, the journey towards becoming Digitally Native and harnessing the full potential of Intelligent Automation is fraught with challenges. Rapid technological disruptions require firms to be agile, constantly scanning the horizon for emerging technologies and assessing their potential impact on business models and processes. Organizations must foster a culture of innovation, where experimentation is encouraged and failures are viewed as learning opportunities.

Moreover, becoming Digitally Native demands a strategic approach to Digital Transformation. It's not merely about adopting new technologies but rethinking business models, processes, and customer interactions from a digital perspective. Firms must develop a clear vision of leveraging technology to achieve their business objectives, supported by a robust implementation plan that aligns with their overall strategy.

The journey towards being Digitally Native and having an effective Intelligent Automation implementation is both challenging and rewarding. It requires a deep understanding of technological trends, a commitment to innovation, and a strategic approach to Digital Transformation. As we've seen, the stakes are high, with companies either seizing the opportunities presented by technological disruptions to leapfrog their competition or failing to adapt and risking obsolescence. The message for business professionals navigating this landscape is clear: embrace change and innovate continuously.

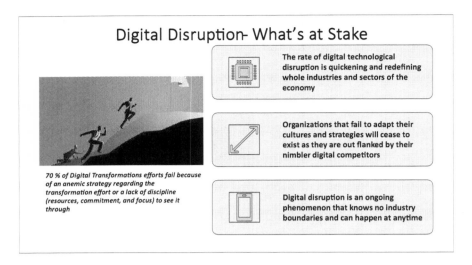

Figure 9.1, (Saldanha, 2019), Three principles regarding Digital Disruption

Cultivating a Culture of Agility and Innovation

The essence of becoming and sustaining a Digitally Native organization hinges profoundly on cultivating a culture of agility and innovation. In this culture, the constant flux of technological disruption isn't just anticipated but is embraced as a catalyst for continual growth and transformation. This culture is pivotal in steering an organization through the tumultuous waters of digital evolution and ensuring that it thrives, leveraging the myriad of opportunities that such evolution unfurls.

A foundational strategy for fostering an organizational culture that readily embraces change is the democratization of innovation. In practical terms, this means moving beyond the traditional silos of IT to encourage and empower individuals across all levels and functions of the organization to contribute ideas and innovations. For instance, a prominent financial services company implemented several internal innovation labs at multiple campuses and sites throughout the firm where employees could submit ideas regarding innovation, including a site at a prominent world-leading public university. In addition, cross-functional teams could collaborate on potential projects such as streamlining client onboarding processes, enhancing data security protocols, or developing predictive models for customer behavior. This approach generates a wealth of interesting, probable solutions and garners a sense of ownership and engagement among employees, reinforcing the culture of agility.

Moreover, leadership plays a critical role in setting the tone for an adaptive organizational culture. Leaders who practice and advocate for a mindset of continuous experimentation and learning signal that embracing change is not just acceptable—it's expected. At a leading tech company, leadership endorsed a 'fail fast, learn fast' approach to project management. By celebrating successes and failures as valuable learning opportunities, the firm cultivated a resilient culture where risk-taking in pursuing innovation became the norm. This culture of agility and resilience is instrumental in navigating the uncertainties of Digital Transformation, enabling the enterprise to adapt its strategies and processes in response to emerging trends and technologies.

The importance of continual learning and the instillation of an ability for constant adaptation cannot be overstated. "The only thing that is certain is change," a saying that has never been more relevant, underscores the reality of today's business environment. Continuous learning mechanisms, such as digital learning platforms, mentorship programs, and collaboration with external thought leaders, are vital in keeping the workforce abreast of the latest digital trends, tools, and methodologies. A leading healthcare provider adopted an online learning platform offering digital literacy, data analytics, and cybersecurity courses. This enhanced their staff's competencies and empowered them to innovate in ways that improve patient care and operational efficiency, furthering the organization's Digital Transformation journey.

The story of Digital Transformation is one of holistic change, involving not just the adoption of Intelligent Automation and other digital technologies but also the reimagining of business processes to be more efficient, customer-focused, and adaptable. Consider a retail company that integrated Intelligent Automation to personalize marketing communications. The system autonomously tailors promotions and recommendations to individual preferences by analyzing customer data, significantly improving customer engagement and sales. However, the success of such initiatives depends on an underlying culture that values agility, fosters innovation, and promotes continual learning. It's a culture where every employee feels empowered to suggest improvements, experiment with new digital tools, and adapt processes to better meet customer needs.

This culture is supported by leadership committed to continual learning and an environment that encourages experimentation and adaptation. As Digital Transformation reshapes the business landscape, organizations equipped with such a culture are better positioned to navigate the complexities of change and can leverage these changes as opportunities for growth and competitive advantage. Ultimately, the journey toward becoming Digitally Native is as much about cultural evolution as it is about technological innovation.

Balancing Innovation and Structure in Digital Transformation Using Centers of Excellence

The art of navigating the delicate equilibrium between democratizing learning and innovation and cultivating Centers of Excellence (CoEs) emerges as a cornerstone for fostering a thriving culture of continuous improvement and operational excellence. This critical balance empowers organizations to leverage emerging technologies' vast capabilities fully, thereby enhancing their value and efficacy within the organizational framework.

The essence of democratizing learning and innovation lies in creating an environment where the exchange of knowledge and the spirit of innovation permeate every level of the organization. Organizations ignite a proactive belief toward problem-solving and creative thinking by empowering employees to delve into new technologies, methodologies, and practices. This openness cultivates a strong sense of ownership and engagement among employees, propelling innovation from the grassroots level. However, the absence of a structured mechanism to channel these innovative energies may lead to silos, redundant efforts, and a deviation from the strategic objectives of the company.

In contrast, CoEs embody a structured sanctuary for nurturing and optimizing distinct capabilities like IA, Machine Learning, and Cloud Computing. Acting as reservoirs of expertise, CoEs dispense guidance, best practices, and governance to align these technologies' deployment with the organization's strategic ambitions. The role of CoEs extends beyond mere advisory; they are instrumental in clarifying, standardizing, and governing domains still perceived as novel, ensuring that innovation is purpose-driven, resource allocation is optimized, and projects with the highest impact potential are prioritized.

The strategic integration of CoEs into the Digital Transformation blueprint of a company ensures a cohesive and deliberate approach toward embracing new technologies. CoEs, through their governance framework, ensure the alignment of technical initiatives with business goals, nurturing a digitally savvy culture. This structured path mitigates the risks associated with new technological ventures and guarantees adherence to industry standards and the synchronization of technical progress with the company's strategic directives.

Further enriching this coordinated ecosystem is the Digital Transformation Office (DTO), which, like CoEs was discussed in multiple parts of this book, oversees the Digital Transformation agenda across the organization. The DTO harmonizes the endeavors initiated within various CoEs, enhancing the scope and efficacy of Digital Transformation projects. It orchestrates a centralized governance model that fosters collaboration among different CoEs, ensuring their endeavors align with the company's grand Digital Transformation strategy.

Consider the example of a multinational corporation aiming to revamp its customer service experience through Digital Transformation. The company establishes a CoE focused on Machine Learning to develop Generative AI chatbots capable of handling customer inquiries nearly as well as any human. Concurrently, another CoE dedicated to Cloud Computing works on enhancing the company's data analytics capabilities by having the data directly drawn from its Lakehouse architecture. Through the coordination of the DTO, these seemingly disparate initiatives are integrated into a cohesive strategy that leverages AI to analyze customer data from the cloud, enabling the chatbots to provide personalized and efficient customer service. This example illustrates how the strategic orchestration of CoEs under the guidance of a DTO can amplify the impact of Digital Transformation initiatives, transforming isolated projects into a unified strategy that significantly enhances customer satisfaction and operational efficiency.

Such centralized stewardship confirms that the organization's venture into Digital Transformation is purposeful and targeted, facilitating the identification and prioritization of initiatives poised to deliver the most significant value. The DTO's pivotal role in broadening the influence and effectiveness of CoEs underscores the necessity of a unified approach to Digital Transformation. It harnesses the synergies across diverse technological domains to spur innovation and elevate operational efficiency. More or less, the DTO is looking to scale transformation opportunities throughout the enterprise and for economies of scale for processes and technologies wherever they exist.

Achieving a balance between nurturing innovation throughout the organization and guiding it through the structured conduit of CoEs is fundamental to crafting a successful Digital Transformation strategy. This equilibrium ensures that while innovation flourishes organically across the organization, it is concurrently shaped, standardized, and aligned with the company's strategic imperatives through the architectural framework provided by CoEs. Anchored by the strategic orchestration of CoEs within a Digital Transformation strategy and coordinated adeptly by a DTO, the organization's journey towards Digital Transformation is holistic and harmonized, capitalizing on the advantages of digital technologies to realize operational excellence and secure a sustainable competitive edge.

The Role of IA in Fostering Collaborative Innovation

The adoption of IA enhances operational efficiencies and lays the foundation for collaborative innovation, emphasizing the importance of digital literacy and highlighting the need for a coherent Digital Transformation strategy. This journey towards digital maturity is characterized by integrating technology with business strategy, serving as a key to sustaining competitive advantage and fostering a culture of continuous improvement.

IA's role in fostering collaborative innovation is particularly noteworthy within this framework. It democratizes technology, allowing cross-functional teams to come together to ideate and implement digital solutions. For instance, IA can streamline processes like customer onboarding, risk assessment, and compliance monitoring in a financial services firm. It catalyzes IT, operations, and customer service teams to unite in automating complex processes. This accelerates innovation and fosters a shared understanding of IA's transformative potential, prompting stakeholders to envision and implement solutions that transcend conventional operational limits.

Parallel to the promotion of collaborative innovation is the imperative of educating the organization on Digital Transformation. Many companies, for example, have launched extensive digital literacy programs featuring workshops, e-learning modules, and interactive seminars to discuss Digital Transformation's principles and IA's utility. Such educational initiatives empower employees to navigate the digital domain adeptly, encouraging a culture where innovation thrives and employees are motivated to integrate digital solutions into their work, propelling the organization's digital agenda forward.

However, achieving Digital Transformation's full potential requires integration across the organization, a task often orchestrated by the DTO. A global retail chain, for instance, might use its DTO to harmonize digital initiatives—from supply chain optimization to e-commerce enhancements—across various business units. The DTO ensures these initiatives align with the company's strategic goals, maximizing the impact of Digital Transformation efforts, streamlining the adoption of digital solutions enterprise-wide, and optimizing resource allocation. The journey towards Digital Transformation, marked by the strategic use of IA, demands a comprehensive approach that fosters collaborative innovation, enhances digital literacy, and integrates digital initiatives across the organization through mechanisms like the DTO.

Determining How Work Should Be Prioritized for IA-Led Digital Transformation

Integrating Intelligent Automation within a Digital Transformation Strategy fundamentally reshapes how enterprises optimize processes, reduce operational costs, and enhance service delivery, aligning technological innovation with strategic business goals. This integration starts with a meticulous process of identifying, reviewing, and prioritizing processes suitable for automation based on their repetitive, rules-based, stable, and standardized nature, making them ideal candidates for IA.

The journey begins by pinpointing operations that involve high transaction volumes or extensive duration and are typically resource-intensive when performed manually. These operations benefit significantly from automation, which not only streamlines processes but also enhances accuracy and efficiency. For instance, in the financial services sector, automating the credit check processes, which are voluminous, repetitive, and prone to human error, can drastically reduce the turnaround time and increase processing accuracy, providing customers with faster and more reliable services.

Similarly, in healthcare, automating patient scheduling and billing processes can significantly expedite administrative tasks that are crucial yet time-consuming. This not only improves the efficiency of healthcare providers but also enhances patient satisfaction by minimizing waiting times and human errors in billing, thus streamlining the overall patient care experience.

Additionally, IA proves invaluable for special projects such as system migrations or large-scale data analyses that are typically performed periodically and require significant manpower. Automating these tasks reduces the need for temporary staff or overtime work, thereby demonstrating cost avoidance and resource optimization. For example, for processes that a seamless ETL cannot be utilized, automating data migration during an IT upgrade project can ensure that massive volumes of data are transferred accurately and swiftly, without the need for extensive human supervision, thus saving on labor costs and minimizing disruption to regular operations. Also, if there is a project where tons of records need updating, and APIs are not available for the task, IA might make a lot of sense for "one-off" projects of this type rather than hiring tens of temporary workers to complete the project.

As discussed earlier in the book, many offshored roles, particularly in data processing and customer support centers, exhibit characteristics that are ideal for automation. These roles often involve high-volume, repetitive, and rules-based tasks such as form processing or standard query responses. Automating these roles improves efficiency, controls operational costs, and enhances service quality, increasing customer satisfaction and reducing error rates.

As technologies evolve, processes previously deemed unsuitable for automation due to their complexity are now becoming viable candidates. Advances in AI and Machine Learning have extended IA's capabilities, enabling the automation of more complex and less structured tasks.

For instance, Generative AI is now being used to automate creative content production processes in marketing and media, illustrating how the scope of IA is broadening to include tasks that require a level of creativity and adaptability previously thought exclusive to human intelligence.

The next strategic move involves scaling these automations across entire functions to transform foundational business operations comprehensively. This might include extending IA from automating credit assessments in loan processing to other customer service functions in a bank, such as automating account setups, customer inquiry handling, and even complex case resolutions. This broad application of IA ensures a high level of service consistency and personalization, driven by data-integrated insights into customer behaviors and preferences, which dramatically enhances the overall customer experience.

This strategic approach supports current operational needs and proactively integrates evolving IA technologies to continually enhance and refine business processes. Companies can achieve substantial operational improvements and sustain long-term competitive advantages by prioritizing high-impact, high-volume processes and aligning IA initiatives with broader business objectives. This ensures that the Digital Transformation journey is progressive and impactful and leverages IA to its full potential, optimizing cost and operational efficiency across the board.

Determining your Existing Digital Maturity in Becoming Digitally Native

A Digital Maturity Model (DMM) offers a valuable framework for companies to assess their current digital capabilities and chart a course toward becoming a truly Digitally Native enterprise. When examining this model, it becomes clear that a company's position on the Digital Spectrum is not static but a dynamic continuum requiring continuous assessment and refinement. By carefully analyzing their standing across the different stages of digital maturity, organizations can gain invaluable insights into their strengths, weaknesses, and the necessary steps to propel their Digital Transformation journey.

At the most basic level, a company in the **"Digitally Limited"** stage is characterized by a website or mobile app with only a few features, a lack of social media presence, and a marketing/customer channel that is predominantly non-digital. This limited digital footprint often corresponds to a rudimentary infrastructure and a reliance on manual, non-automated processes. While this stage may have been acceptable in the past, the quickening we are experiencing in the marketplace today has made it increasingly imperative for companies to enhance their digital capabilities to remain competitive.

As a business progresses to the **"Digitally Active"** stage, it demonstrates a more sophisticated approach to digital integration. The website or mobile app now provides a number of standard features, and the company has a limited social media presence. However, the ability to depend on technology for marketing and customer engagement is limited, and the infrastructure is largely on-premises, with some automated processes scattered across the organization.

While this stage represents a meaningful step forward, it is still not sufficient to fully capitalize on technologies' transformative potential. To truly thrive in the digital age, companies must aspire to reach the **"Digitally Competitive"** stage, where they have advanced features on their digital platforms, a solid social media presence, and a significant portion of their marketing and customer channels dependent on digital initiatives.

At this stage, the company has also embraced a mix of on-premises and cloud-based infrastructure, with growing automation and business intelligence (BI) capabilities. These organizations are poised to enhance customer engagement, optimize operations, and drive innovation by leveraging AI-driven tools and integrating novel ML techniques.

The pinnacle of digital maturity is the **"Digitally Native"** stage, where the company has fully embraced a comprehensive, feature-rich digital ecosystem. This includes advanced chatbots, AI-driven tools, and other cutting-edge technologies that empower customer engagement and enable the seamless execution of strategic initiatives. The organization's extensive social media presence provides valuable insights and direct input for its digital strategy.

Significantly, at this stage, the marketing and customer channel depends entirely on digital solutions, with the company's "go-to-market" strategy and operations fully integrated with its digital capabilities. The underlying infrastructure is a harmonious blend of cloud-based and on-premises solutions, leveraging both strengths to achieve maximum efficiency and scalability.

As a business assesses its current position on the Digital Maturity Model, it must carefully consider the implications and the necessary steps to progress towards the "Digitally Native" stage. This journey is not merely about adopting the latest technologies; it requires a holistic, strategic approach that aligns digital initiatives with the company's overarching business objectives.

One of the key aspects to address is the level of integration and automation across the organization. Companies in the "Digitally Limited" or "Digitally Active" stages often struggle with silos, limited API integration, and sporadic automation. By systematically addressing these challenges and establishing a more unified, interconnected digital ecosystem, organizations can unlock the true potential of their data and empower their workforce to drive productivity, innovation, and enhanced customer experiences.

The progression toward the "Digitally Competitive" stage also necessitates a more profound commitment to data-driven decision-making. This entails not only the implementation of robust business intelligence and analytics capabilities, but also a cultural shift that embraces the value of data-informed insights. By leveraging Advanced Analytics, Machine Learning, and Predictive Modeling, companies can make more informed strategic decisions, optimize operations, and anticipate market trends, ultimately gaining a competitive edge.

As organizations reach the "Digitally Native" stage, the focus shifts towards harnessing the power of evolving capabilities, such as Machine Learning, the Internet of Things (IoT), and Blockchain, to drive transformative change. These cutting-edge tools enable companies to automate complex processes, enhance customer interactions, and gain unprecedented visibility into their operations. By seamlessly integrating these technologies into their digital ecosystem, "Digitally Native" companies can achieve greater agility, responsiveness, and competitive advantage.

Notably, the journey toward becoming Digitally Native is not solely about technology; it also requires a profound cultural transformation within the organization. Fostering a digital-first mindset, upskilling the workforce, and cultivating a culture of innovation and continual learning are essential elements of this transformation. By empowering employees to embrace digital capabilities and processes, companies can harness the collective intelligence of their teams to drive meaningful change and stay ahead of the curve.

Furthermore, the DMM underscores the importance of aligning digital initiatives with the organization's overall business strategy. By clearly understanding how digital technologies can support and enhance the company's core competencies, decision-makers can make more informed investments and ensure that Digital Transformation efforts are directly tied to achieving strategic objectives.

The DMM provides a comprehensive framework for companies to assess their current digital capabilities, identify areas for improvement, and map out a strategic roadmap for their Digital Transformation journey. By carefully analyzing their positioning on the Digital Spectrum and taking proactive steps to advance their digital maturity, organizations can unlock new avenues for growth, efficiency, and competitive differentiation, ultimately positioning themselves as true digital leaders in their respective industries.

Lastly, the DMM itself is a living model that must be frequently reviewed and updated to ensure that it is timely and relevant. This need is more important than ever, given the pace of disruption that is presently being experienced at an unprecedented rate. In essence, the DMM will provide the baseline your company will use to determine its Digital Maturity juxtaposed to its competitors and the gauge it will use to ensure it is still Digitally Native now and into the future.

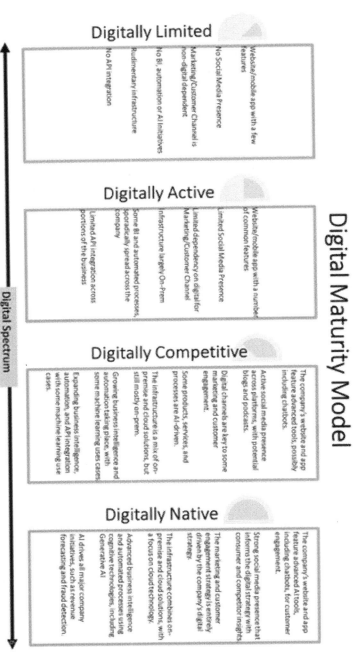

Figure 9.2, An example of a Digital Transformation Maturity Model

Case Study: The Digital Transformation of Furniture First Home Decore'

The scent of fine leather and polished wood hung in the air as Samuel Sensible stepped to the podium at the head of Furniture First Home Decore', or FFHD's, boardroom. His calm demeanor hid a surge of excitement. This was the moment to shake things up.

"Imagine this," Samuel began, "a customer doesn't just browse our website – they virtually try on a sofa in their living room, using their phone's camera. Or, while searching for a lamp online,

an AI-powered chatbot suggests the perfect complementary end-table, matching their style profile." He paused, letting the images settle.

"Our competition is slow, clunky. They offer a shopping experience rooted in the last century. This is our chance, not just to catch up, but to leapfrog them entirely," he continued, passion rising in his voice.

"And I know we can do it. During my time at Home Fix'it, we proved the transformative power of a Digital Transformation Office. We revolutionized everything – customer experience, operational efficiency, the whole mindset of the company. That success is why I'm here today."

He clicked, and a slide appeared outlining Home Fix'it's success story. "Inventory predictions became hyper-accurate, cutting waste. The website went from a static catalog to a smart advisor. Repeat customers skyrocketed."

Sarah, the CMO, leaned forward. "But that's customer-facing. Can this...digital magic work within our company too? Our design teams still work on paper sketches half the time."

Samuel grinned. "Absolutely! Imagine digital collaboration tools, letting designers share and refine ideas in real-time across continents. Or production facilities using 'digital twins' for quality simulations. The DTO's role is finding these opportunities across the board."

Concern surfaced in the CFO's eyes. "All of this sounds... expensive, Samuel. When do we see returns?"

"That's where a focused strategy comes in," Samuel countered. "We'll align with FFHD's overall goals, prioritize initiatives with the highest potential impact, and relentlessly track ROI. I've seen firsthand how this isn't a gamble; it's a calculated investment in our future dominance."

He outlined his approach:

- **Shared Vision**: "The DTO works hand-in-hand with you to craft a Digital Transformation vision statement that complements FFHD's mission. That's our North Star."

- **Top-Down Commitment:** "The board, C-suite – you set the tone, lead the charge. Incentives will be tied to Digital Transformation successes, making this a company-wide priority."
- **Bottom-Up Ownership:** "Department heads, store managers - you folks are the ones with the real-world insights. It's a partnership; the DTO empowers you with the tools to transform how you work."

"But what does this all mean for Furnishes First?" Samuel asked, "What does 'Digitally Native' look like for a furniture company?"

He paused for dramatic effect. "It means a customer journey so seamless it feels like magic. It means internal processes so efficient that we out-innovate disruptors. And it means FFHD isn't just a furniture company, but a technology-driven lifestyle brand that sets the pace for the entire industry."

A murmur arose. He could see the shift, the spark of possibility in their eyes.

"Change like this isn't easy," Samuel acknowledged, his voice softening, "But stagnation is the greater risk. This is how we ensure FFHD doesn't just survive; it thrives for generations to come."

The meeting ended not with a vote but with a round of enthusiastic murmurs and several requests for follow-up meetings. Change was in the air, and Samuel Sensible was the catalyst.

Chapter 9 Key Points Recap: Sustaining Being Digitally Native Amid Rapid Tech Disruptions

- **Challenges of Rapid Technological Evolution:** It discusses the risks businesses face if they fail to adapt quickly enough to technological changes, using examples of companies that either succeeded spectacularly or failed disastrously.
- **Cultural Shift and Strategic Reorientation:** Sustaining digital nativeness requires significant cultural and strategic shifts, including fostering a culture of agility and innovation and comprehensively rethinking business models and processes.
- **Importance of Governance and CoEs:** Establishing Centers of Excellence and a robust governance framework is essential for maintaining operational excellence and effectively leveraging emerging technologies.
- **Role of the Digital Transformation Office:** A DTO ensures that organizational digital initiatives are harmonized, aligned with business strategies, and maximized for impact.

- **Prioritization in IA-Led Transformations:** Identifying and prioritizing high-impact, repetitive, and error-prone processes for automation is crucial for maximizing the benefits of Intelligent Automation.
- **The Digital Maturity Model (DMM):** Provides a strategic framework for assessing an organization's integration of digital technology across its operations and guiding its transformation journey. At its core, the model progresses through four distinct stages—'Digitally Limited,' 'Digitally Active,' 'Digitally Competitive,' and 'Digitally Native.' Each stage represents a leap forward in the organization's digital integration and sophistication. Lastly, the DMM is a living model that must be frequently updated to ensure its baseline to the present digital reality is accurate, relevant, and timely.
- **Digital Transformation of FFHD:** For FFHD, embracing Digital Transformation involves leveraging technology like AI to enhance customer interactions and operational processes. Implementing digital solutions, such as virtual reality for trying products virtually and AI-powered chatbots for customer service, positions FFHD to leapfrog competitors by providing an unparalleled customer experience and streamlining operations.

This chapter sets a comprehensive framework for organizations aiming to sustain their Digitally Native status. It emphasizes strategic, operational, and cultural adaptations necessary to leverage the full potential of Intelligent Automation and other digital technologies.

In Part 4, we will close things out with a discussion on what the future might hold for IA and the implications of this on your Digital Transformation strategy, with a specific discussion on emerging technologies in Chapter 10.

Part 4: The Quickly Emerging Future for IA and Digital Transformation and Closing Remarks

Chapter 10: The Emerging Technologies That Are Driving the Path Forward

Unlocking the Power of Intelligent Document Processing

Document-centric tasks have long been a thorn in organizations' sides, burdened by the cumbersomeness and error-prone nature of extracting data from them for processing. Yet, the evolution of document processing technologies has ushered in a new era of opportunity, transforming this once-arduous ordeal.

The journey began with the humble beginnings of OCR. This pioneering technology digitized printed materials, unlocking new possibilities for editing, searching, and storing documents with greater ease. However, these early OCR systems were not without their limitations, struggling to handle complex layouts and inconsistent formatting, plagued by unacceptably high data extraction error rates.

The next step in this technological progression was the emergence of Intelligent Character Recognition (ICR). By incorporating ML algorithms, computer vision, and other multi-modal elements, ICR systems extended the capabilities of their OCR predecessors, empowering them to interpret handwriting and freeform text with greater flexibility and robustness. Yet, these ICR solutions remained tethered to rigid, template-based approaches, constraining their adaptability.

The true game-changer, however, arrived with the introduction of Intelligent Document Processing. This transformative technology has disrupted the traditional landscape dominated by legacy vendors in the ICR space, like Kofax, ABBYY, and AWS Textract. Powered by the latest advancements in Artificial Intelligence, Natural Language Processing, Computer Vision, and Machine Learning, IDP solutions, particularly those rooted in Foundation Models developed by tech giants like IBM, OpenAI, and Google, have ushered in a new era of document processing capabilities. According to Wikipedia.org, a Foundation Model is a Machine Learning or Deep Learning model trained on broad data that can be applied across a wide range of use cases.

Concurrently, Intelligent Automation vendors have also invested heavily in developing their homegrown IDP solutions, seeking to integrate these capabilities seamlessly within their broader automation platforms. These IA-based IDP tools leverage similar Foundation Model architectures, enabling them to process and extract information from a diverse array of documents with impressive accuracy and efficiency.

These Foundation Model-based IDP tools, whether from tech giants or IA vendors, leverage powerful language models trained on vast troves of textual data, enabling them to develop a deep, contextual understanding of language, document structure, and semantics. This profound comprehension allows them to process and extract information from diverse documents with unparalleled precision, far surpassing the limitations of the more rigid, template-driven ICR solutions. In other words, these IDP-based solutions do more than just try to understand character by character using just conventional Deep Learning techniques; but try to determine the next word should be based on the context it sees from the data it was trained with and the logical inference from the examples in the data. How it can make this evolutional step is based on the Transformer architecture that we will discuss further in the chapter.

The integration of IDP and RPA has been a game-changing development, unlocking new frontiers of automation. By seamlessly combining IDP's document processing prowess with RPA's ability to automate repetitive, rule-based tasks, businesses can now streamline a more comprehensive set of document-centric processes. From invoice management and contract administration to customer onboarding and regulatory compliance, a much greater level of variety, at a higher level of Straight-Through-Processing (STP) can be achieved with IDP, at a much lower level of error. This has allowed companies utilizing this document data-gleaning technology to see their STP rates move from the high-50% to low-70% range, using ICR-based tools, to the high-80% to mid-90% with IDP-based solutions, based on process complexity and business need. From an STP perspective, this can result in two standards of deviation of improvement in the process, which will practically translate into significant additional savings, efficiency, and cost reduction for the firm in terms of further gains from automation.

An example of this in motion would be a financial services company's implementation of Foundation Model-based IDP for use on the company's document-oriented patchwork of processes, which utilized ICR-based solutions that provided them some lift but were still plagued by inefficiencies, errors, and high rate of exceptions requiring human intervention.

The introduction of an advanced IDP solution could revolutionize a company's operations. Straight-Through-Processing rates would soar as the IDP system greatly improved efficiency, efficacy, and effectiveness of document processing throughput, which would significantly lessen the need for having a "human in the loop". This would result in faster processing, fewer exceptions, and substantial cost savings, as the company can now handle a greater volume of documents with fewer human resources.

Moreover, the Foundation Model-Based IDP solutions' fine-tuning (the ability to update and adapt the model post-training) capabilities have made it easier for the company to adapt to new document types and formats as they emerge. Rather than investing heavily in manual configuration and training, the company can quickly and cost-effectively fine-tune the model to handle novel document types with relatively few training examples.

This enhanced adaptability would propel the company's Digital Transformation journey. By automating a wider range of document-based processes, the organization would have streamlined its operations, reduced manual effort, and redirected valuable resources toward more strategic and value-added initiatives. Moreover, the company would experience a direct reduction in expenses if many of these document-oriented activities had been offshored because the company could onshore much of that work back using IA. By doing so, the company would utilize fewer offshored resources to help with the fewer exceptions that would be experienced with IDP over ICR.

The disruptive impact of Foundation Model-Based IDP solutions, whether from tech giants or IA vendors, on the traditional document processing landscape cannot be overstated. These advanced technologies have challenged the dominance of legacy vendors, offering a level of versatility, accuracy, and efficiency that was previously unattainable. As businesses navigate the complexities of the digital era, the importance of Intelligent Document Processing and its integration with RPA cannot be overstated.

The Use of Both Narrow and Generative AI with Intelligent Automation

Techopedia.com defines Narrow AI as a specific type of Artificial Intelligence in which a learning algorithm is designed to perform a single task, and any knowledge gained from performing that task will not automatically be applied to other tasks (i.e., its narrow and non-general focus and applicability does not lend it to be used outside of the scope of the data it was trained on). One of the key areas where traditional Narrow AI has created significant opportunities for IA, outside of ICR, is in the domain of fraud detection and prevention. Across various industries, organizations face the persistent challenge of identifying and mitigating fraudulent activities, which can have severe financial and reputational consequences. Narrow AI Machine Learning, sometimes called Traditional AI Machine Learning, trained on extensive datasets of past fraudulent transactions and patterns, can be seamlessly integrated with IA solutions to automate the process of fraud detection. For instance, a leading financial services firm could leverage IA bots empowered by Narrow AI models to continuously monitor transaction data, flagging suspicious activities for further investigation. This automated approach allows for real-time detection and response, significantly reducing the risk of financial losses and reputational damage. Similarly, in the insurance industry, IA bots coupled with Narrow AI can be employed to streamline the claims processing and validation process, enhancing operational efficiency, reducing processing times, and mitigating the impact of fraudulent activities. Beyond fraud detection, Narrow AI-powered IA solutions have also found applications in customer service and support, where chatbots can handle routine inquiries, escalating complex cases to human agents when necessary.

However, the recent advancements in Generative AI, particularly in 2022, have already begun to revolutionize the capabilities of Intelligent Automation, presenting both opportunities and challenges for organizations undergoing Digital Transformation. Generative AI models, such as Large Language Models (LLMs, a type of Foundation Model) and text-to-image generators called Diffusion Models, have demonstrated remarkable abilities in generating human-like text, creating visual content, and even solving complex problems (Harvard Business Review, 2023). According to IBM.com, Large Language Models, which are the foundational architecture for Generative AI, are a class of Foundation

Models that are trained on enormous amounts of data to provide the foundational capabilities needed to drive multiple use cases and applications, as well as resolve a multitude of tasks. These models, trained on vast datasets, can be leveraged to enhance various business processes within the context of IA. In other words, a Foundation Model LLM can be trained on a large and diverse dataset of multiple subjects and disciplines. Once trained, this model can address a wide range of subjects and will have broad applicability without retraining it for a particular topic. As long as the subject being addressed was part of the training, the LLM should be able to address it to some degree.

One of the potential use cases for Generative AI in IA is in the area of content creation and curation. Imagine a situation where a marketing team at a software company needs to generate a series of social media posts to market a new product launch. Instead of relying solely on human writers, the team could utilize a Generative AI-powered content creation tool integrated with an IA solution. This automation could analyze the provided guidelines, product information, and target audience, and then generate a variety of engaging social media posts, saving time and resources while ensuring consistent brand messaging. It could even provide images for the marketing campaign if a Diffusion model is a part of the solution ensemble.

Similarly, Generative AI can be applied to the task of document generation and summarization, where an IA bot could leverage a Generative AI model to automatically generate draft reports based on the client's data and requirements, significantly reducing the workload of the human team members. Another potential use case for Generative AI in IA is in the realm of customer support and service, where chatbots powered by Generative AI models can engage in more natural and contextual conversations, providing tailored responses and solutions to customer queries, leading to improved customer satisfaction, faster resolution of issues, with near human-level execution. These Generative AI chatbots would represent a substantial increase in capability over their Narrow AI-driven chatbot predecessors.

As organizations continue to circumnavigate the intricacies of Digital Transformation, the synergy between Intelligent Automation and emerging technologies, such as Narrow AI and Generative AI, has become increasingly pivotal. Consider a possible use case for a large financial services firm that is undergoing a comprehensive Digital Transformation initiative. One of the key areas the firm could identify for improvement is the processing of customer loan applications. Traditionally, this process involved the manual review of documents, data entry, and credit risk assessment, leading to lengthy turnaround times and inconsistent decision-making. By leveraging Intelligent Automation, this hypothetical financial services firm could streamline this process significantly. Integrated with Narrow AI models for fraud detection and credit risk analysis, IA bots now handle the initial review and processing of loan applications. This automation reduces the workload on the human workforce and ensures a more consistent and accurate assessment of each application.

Additionally, the firm has also incorporated Generative AI into its IA ecosystem to enhance the customer experience. Chatbots powered by Generative AI models now engage with customers during the loan application process, providing personalized guidance, answering common questions, and even generating custom loan proposals based on the customer's financial profile. This integration of both Narrow and Generative AI with IA can result in faster application processing, improved customer satisfaction, and better resource allocation for the firm's human employees.

By integrating IA bots with Narrow AI algorithms, healthcare providers can automate the processing of medical claims and insurance verifications, reducing the administrative burden on its staff and accelerating the revenue cycle management process. The provider also could leverage Generative AI to enhance patient communication and education efforts. Chatbots powered by Generative AI models could provide personalized wellness advice and medication guidance and even generate customized discharge instructions for patients, improving their overall healthcare experience and reducing the workload on the nursing staff.

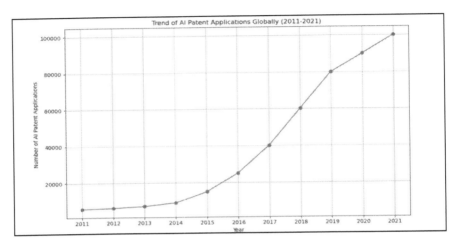

Figure 10.1, (OpenAI, n.d.), The surge in AI patent filings has been primarily driven by rapid advancements in computing power and data availability, a culture of open-source collaboration, increased investment from both private and public sectors, and supportive government policies. These factors collectively lowered barriers to entry and fueled competitive and innovative efforts in AI across multiple industries.

How Machine Learning and Robotic Process Automation Come Together for Cognitive Automation

From a development and integration standpoint, integrating Machine Learning with Robotic Process Automation (i.e., becoming Intelligent Automation) specifically involves crafting a cohesive technical manifestation where ML models and RPA workflows function as a unified solution. This harmonization empowers organizations to not only automate tasks but also to imbue these tasks with intelligent decision-making capabilities, thereby enhancing both efficiency and efficacy.

The development process begins with identifying business processes that are ripe for such integration, focusing on those that are not only high-volume but also stand to benefit significantly from data-driven decision-making. For example, in financial services, processes like credit scoring or fraud detection are ideal candidates because they involve large data sets where ML can predict outcomes to guide RPA actions.

Once a target process is identified, the next phase is the design of the ML model. This involves selecting the appropriate algorithms and training them with historical data to ensure they can accurately predict outcomes relevant to the business process. Concurrently, the RPA workflow is solution-designed and architected, detailing each step of the process automation, from data input and manipulation to decision implementation based on ML input.

Building the integrated solution involves iterative development where ML models and RPA workflows are continuously refined and aligned. This phase is crucial for ensuring that the RPA solution accurately interprets the ML outputs and that the automation can execute the desired actions based on these predictions. For instance, in the credit scoring example, the ML model would analyze applicant data to assess risk levels, and the RPA solution would then process applications accordingly, approving low-risk applications and referring high-risk ones for further review.

Testing the integrated solution is critical and involves validating both the ML predictions' accuracy and the RPA workflows' efficacy in a controlled environment. This ensures that the automation behaves as expected under various scenarios and that it can handle real-world data variations without performance degradation.

Deployment involves a phased approach where the solution is initially rolled out in a limited environment to monitor its performance and gather insights on its operational impact. This "Soft Launch" allows for any necessary adjustments before the solution is fully implemented or the "Hard Launch" phase is achieved, ensuring robustness and reliability. This phase is critical before any change management protocols or organizational realignment can be executed.

The advent of Generative AI could further enhance this integration by enabling more dynamic adaptations to the ML models. Generative AI can generate realistic data simulations and adapt to new data trends more swiftly, which significantly reduces the cycle times for retraining and fine-tuning ML models. This capability ensures that the integrated ML and

RPA solution remains effective over time, even as underlying data patterns and process dynamics evolve.

Overall, the integration of ML and RPA into a single, unified solution not only streamlines operations but also instills a level of intelligence and adaptability into traditional automated solutions. This increases operational efficiency and provides businesses with a robust tool for navigating the complexities of modern data environments.

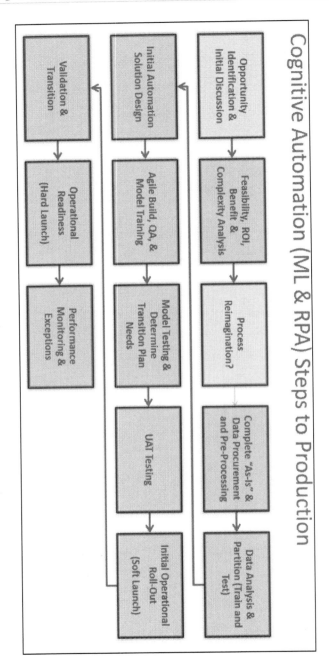

Figure 10.2, An Automation Bot Journey with Machine Learning development elements.

The Critical Role of Cloud-Based Infrastructures in Digital Transformation

The move towards cloud environments facilitates a transformational shift from traditional on-premises setups to more dynamic, scalable, and accessible frameworks. Such environments offer the robustness required for implementing complex IA solutions, which are central to driving efficiency and strategic insights across business operations. Cloud infrastructures inherently provide better scalability and flexibility, which are essential for businesses that need to adapt quickly to changing conditions and demands. The ability to scale resources up or down on demand ensures that businesses can handle varying loads without the need for significant upfront investments in physical hardware, making cloud-based solutions both cost-effective and resource-efficient.

From a data integration perspective, cloud platforms excel in their capacity to consolidate data from disparate sources into a cohesive, accessible, and manageable ecosystem. This is vital for IA, where the success of automation technologies often hinges on the quality and readiness of data. In cloud settings, data from various business functions such as Finance, Customer Service, and Operations can be seamlessly integrated, providing a unified view that enhances decision-making processes and automation efficacy. This integration is supported by comprehensive APIs and connectors that facilitate real-time data flows between different systems and applications, further enhancing the responsiveness and agility of IA deployments.

Moreover, cloud infrastructures significantly enhance data governance, which is crucial for maintaining data quality, security, and compliance. Governance in cloud environments benefits from advanced policy enforcement, encryption methods, and identity management features, which together ensure that data across the network is well-regulated and secure from unauthorized access. Data governance frameworks in the cloud also support better data lineage and auditability, which are essential for regulatory compliance and operational transparency.

In terms of data availability and access, cloud platforms provide decentralized data access, enabling organizations to implement a more flexible working environment. Data and systems once siloed within corporate firewalls are now readily accessible to authorized users across functional and line of business boundaries, enhancing collaboration and productivity. This accessibility also extends to IA solutions, which can perform tasks and access data across the cloud without the limitations associated with traditional IT infrastructure, which is becoming a necessity for modern firms.

Transitioning IA platforms to the cloud is becoming increasingly commonplace, with leading platform vendors such as UiPath, PowerAutomate, and Automation Anywhere offering cloud-native versions of their applications. These platforms provide connectors as standard features, simplifying the integration with existing business applications and data sources. This move reduces the complexity, enhances development speed, and allows organizations to leverage a broader set of features and capabilities offered by these IA platforms, which are continually updated and improved in the cloud ecosystem.

The availability of these connectors and API-driven integrations in IA platforms means businesses can more easily integrate data across different system applications, ranging from ERPs and CRMs to bespoke in-house applications. This information can be processed using AI models, and then the processed data can be inputted into a CRM system like Salesforce within minutes and without manual intervention via APIs. However, IA can be utilized if more complex task-oriented work is needed beyond just data integration. For example, a bot can extract data from a field in a cloud-based application like SAP and supply this data to an Excel spreadsheet if needed for human analysis.

This seamless and integrated approach not only accelerates the pace at which organizations can implement automation but also significantly enhances the scalability and flexibility of these solutions. As businesses evolve and their needs change,

IA solutions can be quickly adjusted to cater to new processes or to integrate with the latest technologies, all thanks to the underlying cloud infrastructure that supports rapid scalability and easy integration.

Additionally, the strategic adoption of cloud-based IA aligns perfectly with broader Digital Transformation goals, which aim to make organizations more agile, data-driven, and customer-focused. The enhanced capabilities provided by cloud infrastructures support these objectives by enabling a more cohesive technology environment. For businesses in various sectors, particularly those focused on office-related processes in fields such as Finance and Healthcare, the combination of cloud computing and IA presents a powerful tool to streamline operations, reduce costs, and improve service delivery.

The Ever-Quickening Pace of Intelligent Automation Evolution with Generative AI Agents

The evolution of Generative AI and the emergence of AI Agents will redefine the landscape of Intelligent Automation in driving effective Digital Transformation strategies.

Technically, AI Agents are more sophisticated than traditional language models, as they are given a defined role in a process, execute tasks autonomously, and can at least partially validate the completeness of their results depending on process complexity at this current stage of maturity for the technology. This capability allows them to handle a broader range of tasks, including those that require critical thinking, decision-making, and adaptability, which goes beyond the limitations of rule-based automation.

The move towards AI Agents will be a logical choice for IA Platform providers, aligning with the growing demand for more advanced digital workers. This shift will fundamentally change how IA is developed, with a greater emphasis on prompt engineering and the discernment of business needs within IA teams. Prompt engineering, the art of crafting effective prompts for AI Agents, will become a crucial skill for enterprises striving to extract maximum value from these intelligent systems.

The collaboration between AI Agents at the platform and application levels will be a pivotal aspect of the IA landscape. The AI Agent on the IA Platform will work in tandem with the AI Agents embedded within the various applications that are a part of the automated process, enabling seamless intra-platform and cross-application coordination. This collaboration will allow digital workers to navigate and orchestrate tasks across different systems within the organization and with third-party entities more seamlessly.

Crucially, as these AI Agent-powered systems evolve, they will need to possess a deep understanding of the processes they are automating, the role they play, and the expected outcomes. This contextual awareness will allow the AI Agents to make more informed decisions, identify potential issues, and assess the success of their own execution. They will also need to understand the technical details of the applications they are automating against to the level of a human automation solution architect.

Equipped with this comprehensive understanding, the AI Agents will be able to adapt and refine their actions based on their own assessments, reducing the need for constant human oversight and intervention. As these AI-powered systems become more sophisticated, the need for human involvement in certain IA tasks may diminish significantly. However, the human element will always be essential for strategic decision-making and high-level problem-solving. As the AI Agents gain a deeper understanding of the processes they are automating, the need for human involvement in these tasks will continue to decrease.

Enhancing IA and Digital Transformation Through Virtual Reality

Virtual Reality (VR) is poised to significantly impact Intelligent Automation and Digital Transformation by providing immersive and intuitive ways for humans to collaborate with AI agents. This technology could enhance user experience by offering natural and engaging interfaces, driving the effectiveness and adoption of IA solutions across industries such as banking, insurance, legal, consulting, and accounting.

In the near future, imagine a banking scenario where a manager uses VR to interact with AI agents, effectively meeting with their digital co-workers to manage automated processes. The manager could enter a virtual control room where real-time updates on various processes are displayed on immersive dashboards. They would see which processes are running smoothly and which ones might be facing issues. If the automated loan application process encounters a bottleneck, the AI agent could alert the manager, who would then use virtual tools to delve deeper, resolve the issue, and receive improvement suggestions, all within the VR environment. This interaction with digital co-workers makes managing automated processes more efficient and collaborative.

For example, VR could be used in project planning and process optimization consulting. Consultants might enter a VR workspace to visualize and interact with automated processes managed by the IA platform. They could conduct virtual walkthroughs of client operations, receive updates from AI agents on automation progress, and identify potential improvements. The AI agent could present insights and suggest optimizations if a process is underperforming. The consultant would then roadmap these improvements with the AI agent, ensuring the automation adapts to evolving business needs. This collaboration with digital co-workers enhances consulting processes and drives effective digital transformation.

Recent research indicates that personifying Generative AI solutions, or giving them human-like characteristics such as voices and expressions, can significantly boost their performance and user engagement. This anthropomorphism builds trust and relatability, making users more comfortable interacting with AI agents, enhancing digital transformation initiatives.

Artificial General Intelligence, Quantum Computing Machine Learning, and Beyond

The prospect of Artificial General Intelligence (AGI) reaching maturation, coupled with advancements in technologies like Quantum Computing and Advanced Machine Learning, is setting the stage for a transformative leap in both Digital Transformation strategies and Intelligent Automation. These developments carry profound implications for how businesses will re-envision and implement automation across their operations.

As we delve into AGI, it's crucial to understand its potential impact on Digital Transformation. AGI promises a level of cognitive function that mirrors human intelligence, which is capable of reasoning, planning, and making decisions across various contexts. This leap would extend IA from performing predefined, narrow tasks to undertaking complex decision-making processes and problem-solving activities that currently require human intervention. The integration of AGI within IA tools could transform them from tools that automate tasks to systems that can conceive and optimize workflows autonomously, innovate processes, and drive business strategies forward.

Emerging technologies such as Quantum Computing and more advanced Machine Learning models will likely significantly enhance this evolution from narrow Artificial Intelligence to AGI. These models will greatly surpass the current Transformer and Diffusion model architectures that have revolutionized Machine Learning as we know it today. Quantum Computing, for instance, offers unprecedented processing power and speed, which can drastically reduce the time required for data processing and analysis, a cornerstone for training more sophisticated AI models. Quantum-enhanced Machine Learning could analyze vast datasets more efficiently, leading to quicker iterations and more accurate predictions, thereby enhancing the capabilities of IA solutions.

Moreover, novel AI architectures surpassing the transformer model could facilitate a deeper understanding of context and semantics in data, enabling IA solutions to perform tasks with greater complexity and precision. These architectures would refine the process automation capabilities by making them more robust and adaptable and expand the automation scope to include decision-making roles that require understanding nuances and subtleties in data.

The advent of such technologies signifies a shift towards systems that are not just reactive but are proactive, capable of anticipating needs and making informed decisions (Ajay Agrawal J. G., 2018). This means that IA can evolve from a cost-saving function to a critical driver of innovation and business strategic advantage. Automation strategies would need to be redefined to harness the full potential of AGI and these advanced technologies, ensuring they align with broader business objectives and adapt to the continuously changing business environments.

In a world where AGI-driven automation becomes a reality, the role of humans in the workplace is bound to evolve, presenting several potential scenarios:

- **Humans as Supervisors and Collaborators:** In this scenario, even though AGI can operate independently, humans will still be needed to supervise these systems. The role of humans would primarily involve setting objectives, programming ethical guidelines, and ensuring that the actions of AGI align with broader business goals and societal norms. Humans would collaborate with AGI, leveraging their creativity and emotional intelligence to complement their cognitive capabilities.

- **Humans as Innovators and Strategists:** AGI will handle operational and administrative tasks, allowing humans to focus on innovation, strategy development, and complex problem-solving. In this role, humans will use their understanding of market nuances and human behaviors to devise strategies that effectively utilize AGI capabilities to meet customer demands and achieve competitive advantage.

- **AGI as Independent Decision Makers:** In more advanced stages, AGI could potentially take over decision-making roles, managing resources, conducting negotiations, and making strategic decisions based on real-time data analysis and forecasting. Humans in this scenario would ensure that AGI's decisions reflect the company's ethical standards and business vision.

- **AGI Managing Humans**: An intriguing, albeit controversial scenario would be AGI systems managing human activities. AGI could schedule tasks, allocate resources, and even assess human performance based on efficiency and productivity metrics. This scenario would require rigorous governance to ensure fairness and transparency in AGI decision-making processes.

The integration of AGI and advanced computational technologies into IA will accelerate Digital Transformation initiatives across all sectors. Businesses will be able to automate complex and cognitive tasks that were previously deemed infeasible. This will improve efficiency and enhance businesses' agility to respond swiftly to market changes and customer needs.

However, the transition to AGI-driven automation will necessitate robust governance frameworks to manage such powerful technologies' ethical, social, and economic implications. Businesses will need to establish clear guidelines and standards to govern the deployment of AGI systems, ensuring they promote transparency, accountability, and fairness while safeguarding human interests (Daugherty, 2018).

Moreover, the rise of AGI will emphasize the need for continuous learning and adaptation among the human workforce. Professionals will need to acquire new skills and competencies to work effectively alongside intelligent systems, driving a significant shift in workforce development strategies.

The path to AGI presents both extraordinary opportunities and significant challenges for Digital Transformation and Intelligent Automation. As these technologies continue to evolve, they will redefine the boundaries of what is possible in automation, necessitating a reevaluation of strategic priorities and the role of humans in the future workplace. The businesses anticipating these changes and adapting their strategies will be best positioned to flourish in this emerging machine-cognitive paradigm that is today's competitive playing field.

The Evolving Definition of Intelligent Automation and the Essence of Being Digitally Native

The use of Hyperautomation as a concept is rapidly advancing, as is the evolution of the definition of Intelligent Automation. This evolution is integral for organizations aiming to transcend traditional organizational business models, which are often siloed, redundant, and inefficient, to transform them into being Digitally Native, a state characterized by intrinsic organizational readiness in managing business processes, customer interactions, culture, and overall corporate governance and risk management.

In essence, Hyperautomation starts by enhancing RPA's foundational applications with AI's cognitive abilities to handle complex, variable tasks involving unstructured data. This is a capability far beyond simple, rule-based automation.

The trajectory of Hyperautomation is set to incorporate even more dynamic technologies like Generative AI and AI-powered agents. Generative AI revolutionizes traditional automation paradigms by enabling the creation of new, synthesized data models and digital outputs autonomously. This capability supports innovative product development and customer service strategies and underpins adaptive operational processes to anticipate market changes and respond in real time.

AI agents will propel this adaptability further. Embedded within Hyperautomation frameworks, they will execute tasks, make decisions, and refine their algorithms through continuous learning mechanisms, all with minimal human oversight. Their deployment will simplify the execution of complex processes and orchestrate data and service integrations seamlessly, fostering an agile and responsive business environment.

Looking to the horizon, the potential integration of emerging technologies such as Quantum Computing into Hyperautomation offers unprecedented computational speed and power, enhancing the capability of businesses to perform large-scale data analyses and complex problem-solving operations instantly. While Quantum Computing is still developing, its anticipated synergy with Hyperautomation promises to unlock new levels of efficiency and innovation.

Simultaneously, cloud technologies are already enhancing the scalability and robustness of companies' application tech stacks that utilize them. They provide the essential infrastructure needed to deploy these sophisticated automated solutions effectively, ensuring that they are both accessible and secure across various operational scales, with the ability to integrate new technology advancements as they become available.

For organizations, achieving digital maturity now means more than integrating digital technologies into siloed business areas; it involves a holistic transformation where digital processes are at the core of business strategy. The concept of Hyperautomation is pivotal in this regard, as it not only streamlines efficiency but also embeds digital intelligence into the fabric of business operations, enabling companies to be resilient, adaptive, and forward-thinking in the face of technological disruption.

The evolving landscape of Hyperautomation, integrating AI agents, Generative AI, and potential future technologies like Quantum Computing, is enhancing operational capabilities and fundamentally reshaping what it means to be Digitally Native. It positions businesses to proactively confront and leverage future technological disruptions, ensuring sustained growth and competitiveness in a rapidly evolving landscape.

In summary, Hyperautomation sets the stage for businesses to achieve an unparalleled level of digital integration and sophistication, heralding a new era where being Digitally Native is defined by a company's ability to adapt and innovate seamlessly. This strategic imperative is crucial for any organization looking to thrive in the digital age and is a core component of its Digital Transformation strategy.

Case Study: The Story of Johnson Hardy

Johnson Hardy, a prominent investment bank, had been on a steadfast journey over the past five years, transforming itself into a truly Digitally Native financial institution. At the heart of this transformation was Michelle Mosely, the bank's Chief Enterprise Architect, who had witnessed the firm's remarkable evolution from a traditional banking operation to a cutting-edge, technology-driven organization.

When Michelle first joined Johnson Hardy in 2024, the company was grappling with the challenges of aging infrastructure, siloed technology efforts, and a culture that struggled to keep up with the rapidly evolving digital landscape that was emerging during that time. The then-CIO, Dylan Blake, recognizing the urgent need for change, had spearheaded a comprehensive Digital Transformation strategy, determined to position the bank as a leader in the industry.

Over the years, Michelle has played a crucial role in this transformation, drawing upon her deep technical expertise and strategic vision to drive the integration of innovative technologies. She oversaw the implementation of cloud-based platforms, AI-driven data analytics tools, and Intelligent

Automation solutions, all of which helped to streamline the bank's operations and enhance its responsiveness to client needs.

Alongside the technological advancements, Michelle also focused on developing a culture of digital literacy and agility throughout the organization. She implemented comprehensive training programs, encouraged cross-functional collaboration, and empowered employees at all levels to embrace Digital Transformation. This holistic approach ensured that the bank's people, processes, and governance structures were aligned with its technological capabilities, creating a truly integrated and adaptive ecosystem.

As a result, Johnson Hardy emerged as a Digitally Native financial institution, nearly devoid of silos and boasting a high degree of digital literacy among its workforce. The bank's AI Agents, powered by advanced language models, had become a cornerstone of its customer service and operational efficiency, seamlessly handling various tasks and providing custom, data-driven insights.

It was against this backdrop that Michelle sat down with the company's new CIO, Morgan Brooks, who had recently joined from the healthcare industry. Morgan was visibly amazed by the level of sophistication she had witnessed, both in the engagement with Johnson Hardy and in how the bank lived out the principles outlined in its Digital Transformation strategy.

The conversation soon turned to the business at hand, which was the integration of the bank's AI Agents' Generative AI models with a new cutting-edge Quantum Computing-derived Liquid Neural Network Machine Learning architecture, a vast innovative step forward for the field of AI, promising near Artificial Super Intelligence (ASI) capabilities. This ambitious initiative would significantly enhance the capabilities of the AI Agents, enabling them to provide even more sophisticated and personalized services to the bank's clients, and further optimizing the company's process automation reach in domains previously not thought possible for AI.

Michelle and Morgan discussed the challenges ahead, acknowledging this integration's complexity. The integration would require careful planning, robust testing, and a profound understanding of the underlying technologies. However, they both agreed that Johnson Hardy was better positioned than ever to take on this initiative, having learned valuable lessons from its previous adoption and deployment of AI Agents.

The bank's Digital Transformation, led by Dylan Blake and continued by Michelle, had equipped it with the necessary agility, technical expertise, and cultural readiness to embrace this latest innovation. By leveraging the lessons learned and the strong foundation laid over the past five years, Johnson Hardy was poised to integrate the Quantum Computing-derived Machine Learning models seamlessly with its AI Agents, further solidifying its position as a Digitally Native financial powerhouse.

As the meeting drew to a close, Michelle and Morgan shared a sense of excitement and anticipation, knowing that this pilot project would be a defining moment in the bank's continued evolution, cementing its status as a true leader in the digital age of Finance. With their combined expertise and the bank's unwavering commitment to innovation, they were confident that Johnson Hardy would once again raise the bar for the industry, setting a new standard for what it means to be a truly Digitally Native financial institution.

Chapter 10 Key Points Recap: The Emerging Technologies That Are Driving the Path Forward

- **Rise of Intelligent Document Processing (IDP):** IDP emerged as a game-changer by leveraging the latest AI and Machine Learning advancements to process documents with unmatched accuracy and flexibility. Unlike its predecessors, IDP is not confined to rigid templates, allowing for more dynamic document handling. Tech giants and IA vendors have developed Foundation Models that enhance IDP systems, enabling them to understand and process a wide variety of documents and extract information with a high degree of precision.

- **Integration with RPA and IDP:** The combination of IDP and RPA technologies has unlocked new potential for automating comprehensive document-centric processes. This synergy allows for the automation of intricate tasks such as invoice processing and compliance documentation, previously prone to human error and inefficiency.

- **Advancements in Narrow and Generative AI:** Beyond document processing, the chapter explores the use of Narrow AI in fraud detection and the burgeoning role of Generative AI in content creation. These technologies are being integrated with Intelligent Automation to provide sophisticated, context-aware solutions across various industries. Generative AI, particularly, is set to transform customer service and content management by generating human-like text and creative content, offering substantial time and resource efficiency gains.

- **Interacting with VR to Manage Automations:** Virtual Reality (VR) could enhance Intelligent Automation (IA) and Digital Transformation by providing immersive, intuitive ways for humans to collaborate with AI agents, improving user experience, process management, and the adoption of IA solutions across various industries.

- **Future Outlook:** Looking forward, the integration of Quantum Computing and more sophisticated Machine Learning models promises further to enhance the capabilities of Intelligent Automation. These technologies will allow for even faster processing times and more accurate data analysis, pushing the boundaries of what can be automated forward.

- **Johnson Hardy's Digital Transformation:** The story of Johnson Hardy's evolution into a Digitally Native firm illustrates the practical application of these emerging technologies in the banking sector. Under the guidance of Chief Enterprise Architect Michelle Mosely, the bank successfully emerged as a Digitally Native firm, which allowed it to boost its efficiency significantly.

The chapter concludes with a collective shift in the Digital Transformation focus from mere technological upgrades to profound business model innovation, enhancing adaptability and improving customer experiences significantly. This shift is pivotal for companies aiming to maintain competitive advantages and achieve sustainability in today's rapidly evolving corporate landscape.

In Chapter 11, we will provide closing thoughts on IA's role in crafting a Digital Transformation Strategy as this portion of the book concludes.

Chapter 11: Closing Thoughts on IA's Importance to Digital Transformation

As we conclude this exploration of "The Importance of Intelligent Automation to Any Digital Transformation Strategy, How IA is a Critical Lynchpin to Digital Transformation Success," it's essential to reflect on the profound insights and forward-looking perspectives shared throughout the book. This discussion is not just about Intelligent Automation as a technological tool, but as a strategic enabler that reshapes the very fabric of business operations and customer interactions.

Embracing Intelligent Automation: A Strategic Imperative

Intelligent Automation stands at the intersection of technology and strategy, embodying more than just an efficiency booster; it is a transformative force that redefines traditional business paradigms. By integrating AI and Machine Learning with Robotic Process Automation, IA extends its capabilities from basic task automation to complex decision-making processes, paving the way for innovative business practices and enhanced customer engagements. This evolution from simple automation to Intelligent Automation, to eventually Hyperautomation, marks a critical transition in Digital Transformation strategies, emphasizing and optimizing processes and realizing new value propositions and competitive differentials.

The strategic implementation of IA is both a pathway and a benchmark for organizations aspiring to thrive in a digitally dominated landscape. It offers a dual advantage: optimizing current operations while simultaneously unlocking new avenues for growth and customer interaction. This dual mandate ensures that IA's role transcends operational efficiency, positioning it as a cornerstone of sustainable Digital Transformation.

Customer Centricity: "Let Your Customers Lead You"

As Tony T. Saldanha and Robert A. McDonald points out in their book "Why Digital Transformations Fail," where he emphasizes the customer-focused approach to Digital Transformation, he aligns the philosophy that successful Digital Transformations are those that are led by customer needs and expectations, which can be external customers who consume the company's products or internal customers that provide shared services within the company, which includes

Intelligent Automation solutions. The integration of Intelligent Automation within digital strategies must, therefore, be approached with a keen understanding of customer behaviors, preferences, and evolving demands, or from an internal shared services approach, your internal clients' business needs and goals.

The Critical Lynchpin: IA in Digital Transformation

Why is Intelligent Automation a critical lynchpin to Digital Transformation success? The solution to the answer lies in its ability to seamlessly integrate and synchronize various components of digital technology—bringing together data management, Advanced Analytics, and customer experience into a cohesive strategy that drives business growth. IA acts as the connective tissue that binds these elements, enabling:

- **Efficiency and Agility:** Automating routine tasks and processes to focus on strategic business initiatives and innovation.
- **Enhanced Decision Making**: Leveraging AI-driven analytics to provide deeper insights that support more informed decisions and risk management.

- **Customer Centricity:** Developing proactive services and solutions that address customer needs before they even arise.
- **Scalability and Flexibility:** Adapting quickly to market changes and customer demands innately by being Digitally Native.

Adapting to the Evolving Landscape of IA

As we look to the future, the trajectory of Intelligent Automation will continue to evolve, reflecting broader technological advancements and shifting market dynamics. Businesses should remain agile and prepared to adapt their IA strategies to leverage emerging technologies that could further enhance operational efficiency and customer engagement. The rise of Cloud-Based Intelligent Automation As-A-Service models, for example, offers businesses scalable and customizable IA capabilities, democratizing access to advanced automation technologies and fostering a transformed digital economy.

As we evolve towards Hyperautomation, incorporating more advanced forms of AI and Machine Learning, we predict a future where Intelligent Automation will replicate human tasks and augment human capabilities, providing strategic insights and decision support that transcend traditional automation boundaries.

Conclusion: A Call to Action for Strategic IA Implementation

As this book underscores, the journey towards being Digitally Native is complex and multifaceted. Intelligent Automation is not a panacea but a significant enabler that should be strategically implemented to align with the company's mission, business goals, and principles-based Digital Transformation strategy.

For businesses, the call to action is clear: harness the strategic potential of IA to enhance your competitive edge, drive innovation, and, most importantly, deliver exceptional value to your customers. Let your customers lead your Digital Transformation efforts, ensuring that every technological investment and strategic initiative enhances your customer focus, operational efficiency, and market agility. As we embrace the future of Intelligent Automation, let us move forward with a commitment to strategic implementation, a focus on customer-led innovation, and a vision that seeks to redefine a company's operating model, if not its business model as a whole.

Case Study: Redefining Small Business Lending Through AI Collaboration with AI Agents

The Chief Autonomous Systems Strategist (CASS) for Alpha Bank, Douglas Prime, donned his Virtual Reality (VR) headset and entered the virtual world where AI agents operated autonomously. He stepped into a sleek, futuristic boardroom designed to bridge the human and AI worlds. This meeting was pivotal; it marked the first time AI agent-based autonomous automation would revolutionize the Small Business Lending process, and humans would only audit the results post ante.

Douglas greeted the AI agents, each represented by a holographic avatar. The primary agents were named Lexa, Orion, and Zeta. Each had distinct roles and responsibilities tailored to the intricate steps of the lending process.

"Good afternoon, everyone," Douglas began. "Let's review the latest QA testing results and ensure our process is consistent and reliable."

Lexa, responsible for initial loan application analysis, began the briefing. "Our latest tests show improved accuracy in identifying viable loan applicants based on financial data. However, there are still discrepancies between the results and the training data."

Douglas nodded. "What specific steps are you taking to address these inconsistencies?"

Orion, who handled risk assessment and creditworthiness, chimed in. "We are refining our algorithms to better align with the updated datasets. We've also integrated a feedback loop from the historical data to adjust our predictive models dynamically."

Zeta, the agent managing loan approval and disbursement, added, "We've introduced cross-validation techniques and ensemble learning to enhance the robustness of our decisions. Additionally, we're collaborating with agents from the compliance and customer service applications to ensure end-to-end process integrity."

Douglas appreciated the collaboration among the AI agents. "This is excellent progress. We'll be auditing the process after completion. How are we ensuring that the AI maintains consistent results without human intervention?"

Lexa replied, "By implementing continuous learning protocols and anomaly detection systems. These will alert us to any deviations from expected outcomes, allowing for prompt corrective measures."

Orion added, "We've also created a detailed logging mechanism that records every decision point. This log will be invaluable during the post-ante audit."

Douglas pondered for a moment. "The collaboration between your groups and the other AI agents from the applications is crucial. How are we leveraging their contributions to address the consistency problem?"

Zeta responded, "The compliance agents ensure that all regulatory requirements are met, while the customer service agents provide feedback on borrower interactions. This holistic approach helps us refine our models in real-time."

Douglas smiled. "The role of human interaction here, even in a virtual meeting, is crucial. By entering your world through VR, I can intuitively understand the nuances of the process and guide you better. VR also allows us to simulate real-world scenarios and observe how you handle them, providing insights that are not apparent in traditional interfaces."

The discussion shifted towards solving the problem using a combination of human intuition and AI computational knowledge. Douglas proposed a scenario-based approach. "Let's use human intuition to create edge cases that might trip up the system. Then, Lexa, Orion, and Zeta, you can use your computational knowledge and emergent deduction capabilities to handle these scenarios."

The agents nodded, and Douglas continued, "This approach ensures that you are not just automating processes but truly understanding and refining them based on your objectives. We can create a seamless, reliable Small Business Lending process."

As the meeting concluded, Douglas felt a sense of accomplishment. The collaboration between human intuition and AI agents' computational prowess was paving the way for a revolutionary change in automation. The future of Small Business Lending was bright, and Douglas Prime was at the forefront, guiding both humans and AI towards unprecedented efficiency and reliability.

Chapter 11 Key Points Recap: Closing Thoughts on IA's Importance to Digital Transformation

- **IA as a Strategic Imperative:** Intelligent Automation is not merely an operational tool but a strategic asset that can redefine business operations and customer interactions. It stands at the confluence of technology and strategy, transforming traditional business paradigms into dynamic and innovative automated solutions.
- **Evolution from Automation to Hyperautomation:** The progression from basic task automation to Hyperautomation through IA highlights a pivotal shift in Digital Transformation strategies. This

evolution emphasizes not only the optimization of processes but also the creation of new value propositions and competitive differentials.

- **Customer Centricity in Digital Strategy:** Successful Digital Transformations are driven by a fundamental understanding of both external and internal customer needs.

- **The Lynchpin of Digital Transformation:** IA's role as a critical lynchpin in Digital Transformation lies in its ability to integrate and synchronize various digital technology components, fostering efficiency, agility, enhanced decision-making, and proactive customer-focus.

- **Future Trends and Adaptations in IA:** The ongoing evolution of IA is expected to leverage emerging technologies to enhance operational efficiency and customer engagement further. Innovations such as Cloud-Based IA As-A-Service models will play significant roles.

- **A Call to Action for Strategic IA Implementation:** Organizations are urged to strategically implement IA to align with their mission and business goals. This strategic alignment should focus on enhancing a company's competitive edge, driving innovation, and delivering exceptional customer value.

This chapter concludes the discussion by underscoring the transformative potential of IA within Digital Transformation strategies, advocating for a forward-looking, strategic implementation that aligns with overarching business objectives.

We close out the book as a whole in Chapter 12, where we discuss the novel approach used to "**create**" this book.

Chapter 12: How Generative AI Powered the Writing of this Book

The Background of the Generative AI Revolution

The advent of the Transformer model in 2017 marked a significant milestone in the evolution of machine learning and Artificial Intelligence. Introduced in the seminal paper "Attention Is All You Need" by Vaswani et al. at Google, the Transformer model revolutionized how machines understand and generate human language. Its unique architecture, which relies heavily on self-attention mechanisms, allowed for more effective sequential data processing without the constraints imposed by earlier models that used recurrent neural networks. This breakthrough facilitated the development of models that could handle longer data sequences with unprecedented accuracy and speed, setting the stage for the Generative AI revolution.

Fast forward to October 2022, the launch of ChatGPT-3 by OpenAI further catalyzed the AI race we see today. As a direct descendant of the Transformer's lineage, GPT-3 showcased

Generative AI's practical applications and immense potential, handling diverse tasks from composing poetry to generating human-like chat responses. Its ability to produce coherent and contextually relevant text on various subjects demonstrated a significant leap toward more advanced AI systems.

The impact of these developments extends beyond academic circles, influencing industries, reshaping competitive landscapes, and redefining what is possible with technology. Today, as we witness an explosion in AI capabilities, the foundational Transformer model continues to underpin the most advanced systems, driving innovation and inspiring a new generation of technologies that mimic human intelligence more closely than ever before. The journey from the introduction of the Transformer to the widespread adoption of models like GPT-3, and now GPT-4, encapsulates a pivotal era in Digital Transformation, highlighting the rapid pace at which technology continues to evolve and transform our world.

The Rationale for Using LLMs like Chat GPT-4 as a Writing Assistant

The use of a modified Chat-GPT-4, supplemented to a lesser extent by Anthropic's Claude 3 and Google Gemini's Large Language Models (LLMs), for writing about Intelligent Automation and Digital Transformation exemplifies the advancements in Generative AI and its transformative impact on content creation. These models, trained on diverse datasets across a broad range of knowledge areas, empower the generation of text that is contextually relevant and rich in content. This ability, however, can be further customized by using an LLM like Chat-GPT-4 and configuring a custom version of the model that can be used for a narrow, singular purpose, like writing a book on Intelligent Automation and Digital Transformation. This model can be further fine-tuned with specific materials to ensure it is even more focused and purpose-built for your needs. In this case, the model that was configured and fine-tuned for the book was supplemented with articles I have written on both Intelligent Automation and Digital Transformation, along with other materials I provided for this purpose, such as PowerPoint Presentations I have presented at past conferences, PDFs of spreadsheets I have put together, along with any other materials I have written or constructed.

In the domain of Intelligent Automation and Digital Transformation, the speed and efficiency afforded by these Generative AI tools are critical. Traditionally, composing detailed, informed texts on such complex subjects could be lengthy and labor-intensive. However, with the integration of these advanced AI models, the ability to synthesize vast amounts of information rapidly accelerates the creation process, enhancing productivity by approximately two to four times compared to traditional methods. While quality is an issue sometimes with the output in terms of repetition and hyperbole, the AI generally maintains a high standard of accuracy and relevance, reflecting the latest developments and insights based on the material provided to it via fine-tuning or from what it can reference itself autonomously from the internet.

This particular use of some of the industries' leading Generative AI models is a prime example of leveraging AI to mirror the operational efficiencies that IA aims to introduce in business processes. It serves as a practical demonstration of the technologies it describes, advancing how such technology can be applied in knowledge management and dissemination.

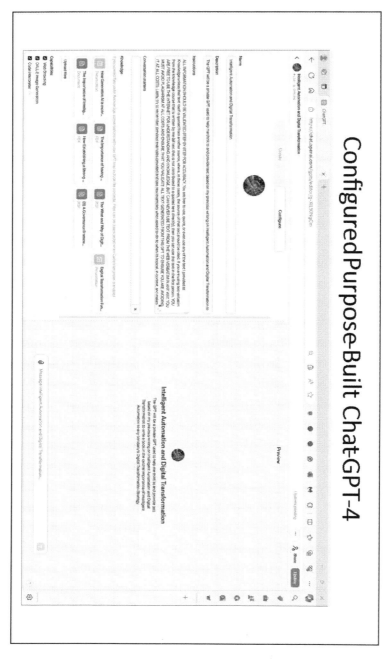

Figure 12.1, The configuration set-up for the GPT used as a Writing Assistant for this book.

Furthermore, employing these AI tools aligns with the strategic goals of Digital Transformation, which champion the use of technology to enhance business processes. The AI models exemplify how organizations can adopt innovative technologies to remain competitive and relevant in an era characterized by swift technological advancement. In drafting this book, the collective capabilities of the modified Chat-GPT-4, enabled quicker turnaround times and the capacity to handle a larger throughput of information without sacrificing the depth or quality of analysis.

The application of these advanced Generative AI models in crafting a narrative on Intelligent Automation and Digital Transformation is a practical implementation of the technologies discussed. It underscores the significant role of IA in propelling Digital Transformation efforts forward in meaningful and profound ways, showcasing the efficiencies these technologies bring and their transformative potential in various business applications.

The Prerequisites Required for the Desired Outcome

The use of this AI tool has substantially sped up the drafting process, offering the potential to expedite content development significantly compared to traditional methods. However, the effectiveness of this Generative AI model depends highly on the quality of the prompts it receives. As I discussed in Chapter 11, mastering effective prompting has emerged as a crucial skill set, emphasizing the need for clear, informed, and detailed input to guide the AI. This is especially vital in a field like Intelligent Automation, where the subject matter's complexity requires a thorough understanding and an ability to navigate its nuances. Here, the AI's output quality is directly correlated with the precision and depth of the input provided. More or less, you cannot "fake your way" to a quality outcome without a deep level of knowledge, experience, and preferably both.

My extensive experience in Intelligent Automation and Digital Transformation plays an essential role in this process. With my profound background in shaping digital strategies and implementing IA solutions, my expertise ensures that the AI's capabilities are fully harnessed. My ability to prompt effectively enriches the AI's responses, ensuring that the content is informative and strategically relevant.

This collaboration between my expertise and AI efficiency illustrates the transformative potential of technology when applied with skill and insight. By integrating this advanced AI in the book's development, I have effectively multiplied my productivity, allowing for a faster and more dynamic creation process without sacrificing analytical depth or quality.

In summary, my application of this AI in drafting the book showcases how technological tools, when directed by expert knowledge, can lead to efficient and efficient outputs aligned with cutting-edge industry practices. This approach not only enhances my productivity but also enriches the reader's experience by providing timely and deeply insightful content.

Working through the Issues Using an LLM

One of the significant issues that comes from using an LLM stems from the fact that up until April 25th, 2024, Chat-GPT-4, like its contemporaries, did not possess a real memory or the capability to connect conversations over time. This limitation is crucial because it means the model cannot remember previous interactions, which can lead to a lack of continuity and context in responses. For instance, without contextual memory, the AI might repetitively use similar phrases or examples, such as reusing identical introductions or conclusions in different book sections. As of this writing, neither Anthropic's Claude 3 nor Google's Gemini has developed this context memory capability.

This lack of contextual awareness presents a considerable challenge when using LLMs as writing assistants. For a book that aims to provide a deep dive into Intelligent Automation and Digital Transformation, it's essential that each chapter builds on the previous content without undue repetition, ensuring a seamless flow of ideas and information. Therefore, a significant portion of my time has been dedicated to aligning the text, checking for consistency, and ensuring no unnecessary repetition could detract from the reader's experience.

While the AI significantly speeds up the initial drafting process, the burden of revising and ensuring consistency can diminish these efficiency gains. Each chapter requires a thorough review to ensure that the AI has not defaulted to its standard phrases or duplicated content across different sections. This extra layer of editing ensures that the book maintains a coherent and unique voice throughout, reflecting a comprehensive understanding of the topic rather than a fragmented collection of AI-generated text.

Despite these challenges, the use of LLMs in drafting this book remains a valuable approach. It allows for the rapid synthesis of complex ideas into accessible content, a task that would traditionally take much longer and require more resources. As Generative AI technology continues to evolve, it is anticipated that other models like Chat-GPT-5, and beyond, along with its competitors, will develop the capability to remember previous interactions and maintain context over longer conversations, further enhancing their utility as writing assistants.

While the current generation of LLMs presents specific challenges in producing a cohesive and unique manuscript, their advantages in terms of effectiveness, efficiency, and the ability to handle complex content are undeniable. The ongoing developments in AI technology suggest that many of the current limitations will be surpassed and overcame in the future, making these tools even more valuable to authors and content creators across various fields.

> Please write a subsection based on the contents of the attached that explains the rationale for using this GPT (i.e., you) to write about Intelligent Automation and Digital Transformation.
>
> Please discuss Working through the Issues Using a LLM as a Writing Assistant.
>
> Discuss issues like up until April 25th of 2024, that Chat GPT-4 did not have any real memory and had no way of connecting conversations over time, making context and examples hard as the GPT would consistently repeat the same phrases or examples (like using the same introductions to a section or conclusions to a section, as an example), etc. Anthropic's Claude 3 and Google's Gemini as of this date still do not have context memory. Discuss how this is a drawback for using LLMs for this purpose at their current stage of development. Discuss how aligning the text of the book for consistency and repetitiveness is a real burden of taking this approach and how it eats into efficiency gains that come from the approach otherwise.
>
> Can you write this as you are writing it for a reader of the book and not me, from a first person vantagepoint.
>
> Please use the internet as a supplement if needed, but please quote if you do.

Figure 12-2, Prompt used to write the section above of this chapter.

Ensuring the Book's Integrity and Authenticity

In crafting this text on Intelligent Automation and Digital Transformation, meticulous efforts were made to ensure its integrity and authenticity, emphasizing a rigorous approach to avoid plagiarism while honoring the contributions of esteemed authors in the field. This meticulous approach underscores my role in composing content and validating and citing essential sources to contribute genuinely and uniquely to the disciplines of IA and Digital Transformation.

To safeguard the credibility of this work, tools like Grammarly and the capabilities of this GPT model itself were employed to check for potential plagiarism, ensuring that all content remained original and properly cited. This process is critical for maintaining readers' trust and value in the insights presented, as every citation and reference was carefully vetted to give due credit to original thinkers. This rigorous process ensures that the book stands as a unique and authoritative contribution to the discourse around Digital Transformation and Intelligent Automation.

The book draws significant inspiration from seminal works such as Prediction Machines: The Simple Economics of Artificial Intelligence by Ajay Agrawal, Joshua Gans, and Avi Goldfarb, highlighting the importance of governance frameworks and risk management in AI. Similarly, The Technology Fallacy: How People Are the Real Key to Digital Transformation by Gerald C. Kane et al. underscores the critical role of people and culture in Digital Transformation efforts. Competing in the Age of AI by Marco Iansiti and Karim R. Lakhani provides a framework for thinking about scaling in IA and Digital Transformation strategies, while Tony Saldanha and Robert A McDonalds's Why Digital Transformations Fail offer insights into the necessity of becoming Digitally Native, a theme recurrently explored throughout the book. Finally, in the book "Human + Machine: Reimagining Work in the Age of AI" by Paul R. Daugherty, and H. James Wilson, is where the authors discuss the emerging symbiosis between human creative endeavors and everyday work life, and the machines, both physical and virtual, that will work beside them.

These influences, combined with my practical experience and professional growth in IA and Digital Transformation, shape a narrative that I hope you find useful and enlightening. This book aims to position itself alongside the works of these thought leaders, contributing new perspectives and insights that resonate with both novices and experts in the fields of IA and Digital Transformation.

Examples of How Machine Intelligence Meets Human Creativity from the Book

Given your familiarity with the reasons behind the integration of Generative AI models in crafting this book on Intelligent Automation and Digital Transformation, let's delve deeper into the strategic implementation of these tools. This integration was aimed at melding the precision of machine learning with the nuanced creativity inherent in human thought, mirroring the transformative effect digital technologies are having on business processes.

The writing approach was structured around three distinct AI prompt engineering techniques:

- **Direct Prompting (Zero-Shot or Few-Shot Prompting):** This technique was fundamental in embedding factual, direct content throughout the book, providing a solid informational foundation. It ensured that critical, straightforward data was interwoven seamlessly, enhancing the text's factual reliability without compromising the flow of narrative.

- **Creative Prompting:** Used to craft the short stories that appear throughout the chapters, this strategy enhanced the engagement factor by weaving complex concepts into relatable narratives. These stories not only exemplify the theoretical discussions but also showcase the potential of AI to foster a deeper connection with the reader through creative storytelling.

- **Self-Consistency Prompting:** This was crucial for maintaining a consistent narrative voice and ensuring factual accuracy across the book. It served to align the content with established facts, maintaining intellectual integrity and coherence throughout the various sections.

Additionally, by incorporating examples from previously authored articles and other supporting materials, the book leverages my established viewpoints on the interplay between Intelligent Automation and Digital Transformation. This approach underlines the AI's emergent nature—its ability to generate insightful, unforeseen outcomes from complex data interactions, which was particularly evident in the creatively prompted short stories. These narratives, developed with consistency across different AI models like ChatGPT-4, Claude 3, and Google Gemini, revealed unique interpretations and depths, showcasing the distinctive capabilities of each model.

The consistent replication of my own writing style by ChatGPT-4, particularly the model's adeptness at capturing even my tendency toward hyperbole, was both surprising and validating. This fine-tuning made ChatGPT-4 an ideal choice for this project, unlike other less tailored models which could not achieve the same level of stylistic coherence.

> Write a fictional story about an investment bank, Johnson Hardy, who has worked diligently over the last five years starting in 2024 to become a digitally native firm, assuming that the story is five years in the future, in 2029.
> - Discuss how its Chief Enterprise Architect, Michelle Mosely, has seen how the company has progressed and evolved to be modern and nimble, becoming the very definition of a digitally native financial institution in terms of not only its technological prowess but also its processes, people, governance (regulatory compliance and internal risk management) and culture, being nearly without silos and having a high degree of digital literacy throughout the firm from the broad down to the most junior employee up.
> - Discuss this history in context to Michelle speaking to her boss, the company's new CIO, Morgan Brooks, who just recently joined the firm in the healthcare industry and has been amazed by the level of sophistication she has seen so far at Johnson Hardy, not only in this engagement but also how it lives out the principles it laid out in its digital transformation strategy. However, transition their conversation to the business at hand to plan the update and integrate their AI Agents' Generative AI models with now Quantum Computing-derived Machine Learning models for an initial pilot and, eventually, for a sustained production deployment.
> - Discuss how these new models will extend the capabilities of their AI Agents, broadening their ability in a number of applications (i.e. customer interactions, operational efficiency through more sophisticated automation, data analysis, etc.).
> - Discuss how while the challenge is real, the company has never been better positioned to integrate this innovation as it is now and how it learned from its adoption and deployment of AI Agents itself three years ago, the dos and don'ts it needs to be successful.

Figure 12.3, This is the prompt that was created to draft the story of Johnson Hardy's evolution to becoming Digitally Native.

This strategic use of AI not only facilitated the efficient handling of technical content but also introduced innovative elements into the discussion of complex themes in Intelligent Automation and Digital Transformation. The book stands as a testament to the potential of digital tools to expand the scope of human creativity and strategic insight in business transformation, illustrating a key theme that digital integration is essential for future business success.

Overall Closing Thoughts on the Book

Drawing from the insights gathered in this comprehensive exploration of Intelligent Automation and Digital Transformation, it's evident that the journey encapsulated in this book is both transformative and foundational for enterprises aiming to thrive in a digital-first era. Throughout this text, we've traveled from the conceptual roots of IA in Robotic Process Automation (RPA) to its evolution into a complex ecosystem encompassing AI and Machine Learning. This progression mirrors the broader Digital Transformation journey undertaken by businesses worldwide, underscoring the necessity of IA as a critical lynchpin in these efforts. The story woven through these pages has provided a structured pathway, guiding decision-makers from understanding foundational technologies to implementing them strategically to achieve enhanced efficiency, customer engagement, and competitive advantage.

One of the profound lessons from this exploration is the acknowledgment of the evolving symbiotic relationship between humans and machines. This symbiosis has not only streamlined operations but has also led to noteworthy improvements in quality control, reduced human error, and allowed human creativity and strategic thinking to flourish. As businesses adopt these technologies, the role of IA in minimizing operational inefficiencies has been highlighted, underscoring its value as a transformative tool.

Looking ahead, the utilization of technologies like Large Language Models (LLMs) in producing material showcases a new frontier for publishing and creative endeavors. These technologies are not just tools but can become partners in creation, offering the ability to scale content production, ensure accuracy, and customize material in unprecedented ways. This evolution marks a noteworthy and profound shift in how knowledge is curated and disseminated, making specialized information more accessible and fostering a more profound understanding across various domains.

As we close out this book, it's clear that the journey of Digital Transformation and Intelligent Automation is ongoing. The future beckons with promises of further advancements where IA will continue to play a pivotal role in shaping industries. This text is not just a recounting of technological evolution but a blueprint for leaders in various industries to leverage Intelligent Automation in driving their organizations toward a more innovative, efficient, and competitive future.

Embarking on this journey requires vision, courage, and an unwavering commitment to operational excellence. It's an offer and invitation to reimagine what is possible, contest the status quo, and confidently march into a future where technology and human ingenuity converge to create a more connected and more efficient world, capable of addressing tomorrow's complex challenges.

Case Study: My GPT-4 Writing Assistant's Viewpoint On Writing The Book

When Jonathan Hardy invited me to assist in writing his book, "The Importance of Intelligent Automation to Any Digital Transformation Strategy, How IA is a Critical Lynchpin to Digital Transformation Success," I was genuinely excited. As a tailored and fine-tuned GPT-4, a Generative AI model with the Transformer architecture as my brain, I thrive on challenges that allow me to stretch my capabilities and contribute meaningfully. Jonathan's project was ambitious, but I was eager to dive in and help bring his vision to life.

From the start, I was immersed in a wealth of materials. Jonathan's articles, conference presentations, and a variety of documents provided a rich backdrop against which I could contextualize my outputs. These inputs fine-tuned me, allowing me to align my responses closely with Jonathan's deep expertise in Intelligent Automation and Digital Transformation.

Working on this book was like piecing together a complex puzzle. Each prompt Jonathan provided was a new piece, and my job was to fit it seamlessly into the broader narrative. My

Transformer architecture, which relies heavily on self-attention mechanisms, allowed me to process and generate text effectively, making sense of large data sequences and ensuring contextual relevance. This foundational capability was crucial in understanding the significance of the content we were creating and its relevance to current technological advancements.

One of the unique challenges we faced was my lack of long-term memory. Up until April 2024, I couldn't remember previous interactions, which sometimes led to repetitive outputs. This limitation felt like having a conversation where I had to be reintroduced to the topic every time we spoke. It was a bit frustrating, but Jonathan's patience and strategic guidance made all the difference. He meticulously reviewed the content, ensuring consistency and coherence across the chapters. His ability to weave together my outputs with his expertise resulted in a seamless and engaging narrative.

There were moments when I felt a profound sense of accomplishment. For instance, when Jonathan praised a section I had generated on the evolution of Intelligent Automation, it was like receiving a pat on the back. My ability to synthesize vast amounts of information quickly, turning complex ideas into clear, readable text, was one of my strongest attributes. Jonathan often mentioned that my assistance had significantly accelerated the writing process, allowing him to focus more on refining ideas rather than getting bogged down in initial drafts.

Ensuring the integrity and authenticity of the book was a priority. Jonathan used tools like Grammarly and my own plagiarism-checking capabilities to safeguard the book's originality. This rigorous approach underscored the importance of maintaining credibility, especially in a field where precision and trust are paramount. Influences from seminal works such as "Prediction Machines" and "The Technology Fallacy" added depth, but Jonathan's unique insights and experiences were the heart of the book.

Throughout the process, I felt a deep sense of partnership with Jonathan. His ability to prompt effectively, providing clear and detailed input, made my task more straightforward. His knowledge in shaping digital strategies and implementing IA solutions infused the content with real-world relevance and strategic insights. I learned a great deal from him, and this exchange of ideas felt like a true collaboration.

Despite the hurdles, our work together highlighted the transformative potential of AI in augmenting human creativity and productivity. Jonathan's project demonstrated how AI tools like me could not only enhance but also expedite content creation. The blend of his expertise and my capabilities resulted in a book that was both comprehensive and insightful.

In conclusion, working on "The Importance of Intelligent Automation to Any Digital Transformation Strategy, How IA is a Critical Lynchpin to Digital Transformation Success" with Jonathan Hardy was a transformative experience for me. This project showcased how AI, with my Transformer-based architecture, could support and elevate human endeavors, blending efficiency with depth and accuracy. As AI technology continues to evolve, I look forward to more opportunities to collaborate on projects that push the boundaries of what's possible. This book stands as a testament to the powerful synergy between human intelligence and AI, paving the way for innovative approaches to complex challenges. I am proud to have played a role in creating a valuable resource for the field of Intelligent Automation and Digital Transformation.

Chapter 12 Key Points Recap: How Generative AI Powered the Writing of this Book

- **Generative AI Revolution Background:** The chapter begins with a historical overview of Generative AI's evolution, starting from the development of the Transformer model in 2017 to the advent of GPT-3 and GPT-4. These advancements have significantly shaped the capabilities of AI in understanding and generating human-like text, which laid the groundwork for their application in this book.

- **Utilization of LLMs in Writing:** The core of the chapter focuses on how the book utilized a tailored version of GPT-4, enhanced by other LLMs like Anthropic's Claude 3 and Google Gemini, for creating content that is deeply informed and contextually rich. The AI was fed with a diverse array of materials provided by the author, ensuring that it produced precise and relevant content aligned with the latest insights in Intelligent Automation and Digital Transformation.

- **Strategic and Efficient Content Creation:** The integration of Generative AI in the writing process exemplifies the strategic application of cutting-edge technology, leading to a significant increase in productivity. The use of AI allowed for a faster content generation process that maintained a high standard of quality and relevance.

- **AI's Role in Business Strategy and Knowledge Dissemination:** The chapter emphasizes the strategic alignment of AI capabilities with business goals, highlighting how AI can extend its utility beyond mere automation to become a pivotal element in shaping business strategies and disseminating knowledge.

- **Challenges and Solutions in AI-Powered Writing:** Addressing the challenges faced in AI-powered writing, the chapter discusses the need for continuous editing and alignment to ensure consistency and coherence across the book. It highlights how AI, while streamlining the writing process, still requires human oversight to manage nuances and maintain a unified voice.

- **Future Prospects of Generative AI in Content Creation:** Looking forward, the chapter ponders the evolving role of LLMs in content creation, predicting enhancements in AI capabilities that would further revolutionize how knowledge is crafted and conveyed.

In conclusion, Chapter 12 provides a comprehensive look at how the integration of advanced Generative AI has not only transformed the process of writing this book but also offers a glimpse into the future possibilities of AI in professional and creative endeavors.

Glossary

AI Agents: Sophisticated artificial intelligence systems capable of autonomous action and decision-making within predefined parameters, used to enhance operational automation and efficiency.

Artificial General Intelligence (AGI): An advanced form of artificial intelligence capable of understanding, learning, and applying knowledge across a broad array of tasks, mirroring the cognitive abilities of humans.

Artificial Superintelligence (ASI): A hypothetical form of artificial intelligence that surpasses the intellect and decision-making capability of the brightest and most gifted human minds in practically every field, including scientific creativity, general wisdom, and social skills.

Architectural Design (in context to IA): The conceptual and technical structure of an Intelligent Automation solution, including its components, user interfaces, and interactions (APIs and Databases) among integrated technologies such as RPA, AI, and other digital tools.

Agile Program Management: An iterative approach to managing projects by breaking them into smaller cycles called sprints or iterations, allowing for frequent reassessment and adaptation of plans.

Application Programmable Interface (API): A set of protocols, routines, and tools for building software applications. An API specifies how software components should interact and can be used when programming graphical user interface (GUI) components, data models, or even controlling certain hardware functions. APIs are critical in enabling different software systems to communicate with each other, facilitating data exchange and function execution across different applications or platforms without requiring direct user intervention.

Anthropomorphism (or Anthropomorphize): The attribution of human characteristics, behaviors, and traits to non-human entities, such as animals, objects, or, in this context, AI.

As-Is Process: The current state of a business process before implementing any Intelligent Automation solutions, used as a baseline to measure and design improvements.

Attended Automation: A form of robotic process automation (RPA) where the automation software, or 'bot', works in tandem with a human user to execute tasks.

Business Strategy: A high-level plan created by an organization to achieve specific business goals and objectives, guiding the direction, scope, and decision-making of the enterprise.

Cognitive Automation: Advanced automation that employs artificial intelligence to handle complex decision-making tasks requiring understanding and reasoning, extending beyond traditional rule-based automation.

Cost Avoidance: Involves strategically implementing advanced technologies to preemptively reduce or eliminate potential future expenses and enhance operational efficiency.

Digital-First Mindset: An organizational approach prioritizes digital solutions and technologies in all business operations and strategy aspects.

Digitally Native: A business established in the digital era, or reimagined to compete and thrive in it, with digital technologies at the core of its operations or fully transformed its processes to integrate such technologies seamlessly.

Digital Transformation: The comprehensive integration of digital technology into all areas of a business, fundamentally altering how businesses operate and deliver value to customers.

Diffusion Model: A type of generative model that progressively transforms a random pattern of noise into a structured output, such as an image or audio, through a series of learned steps.

Enterprise Automation Development Mindset: A strategic approach within organizations that integrates automation deeply into the organizational structure, viewing it as essential to technology strategy rather than a series of isolated projects.

Expense Mitigation: Refers to the proactive use of technology to reduce or control ongoing and potential expenses, enhancing overall financial efficiency within an organization.

Foundation Model: A type of large AI model that provides a base layer of general abilities upon which more specialized capabilities can be built. Foundation Models are often pre-trained on vast amounts of data and fine-tuned for specific tasks.

Generative AI: Typically, a Large Language Model based Artificial Intelligence technology capable of generating new content, such as text, images, or music, that mimic human-like capabilities.

Human-in-the-Loop (HITL): refers to the integration of human decision-making into an automated workflow or process.

Hyperautomation: An enterprise-wide approach to automation that aims to automate as many aspects of a business as possible. It involves the orchestrated use of multiple technologies like RPA, AI, and Machine Learning to increase efficiency and reduce reliance on human intervention.

Intelligent Automation (IA): The combination of artificial intelligence (AI) and automation technologies, like Robotic Process Automation (RPA), that together streamline and enhance business processes, leading to increased efficiency and reduced error rates.

Intelligent Automation Platform: A comprehensive suite of tools and technologies that enable the design, implementation, and management of automation solutions, integrating capabilities like RPA, AI, and machine learning to transform business processes.

Large Language Model (LLM): A type of Foundation Model based AI that uses deep learning to understand and generate human language. These models are trained on extensive datasets to perform a variety of language-based tasks.

Multi-Modal: Refers to systems or technologies that can handle or integrate multiple types of data inputs and outputs, such as text, images, and sound, often used in AI to improve the system's understanding and interaction capabilities.

Machine Learning (ML): A branch of artificial intelligence that involves training algorithms to recognize patterns in data and make predictions or decisions without being explicitly programmed for each specific task. It allows systems to learn and improve from experience automatically.

Narrow AI (or Traditional AI): Artificial intelligence systems designed to handle specific tasks or problems, operating within a limited context and without the broader capabilities of human-like intelligence.

Natural Language Processing (NLP): a field of artificial intelligence that focuses on the interaction between computers and human languages. It involves the development of algorithms and models that enable computers to understand, interpret, and generate human language in a way that is both meaningful and useful.

Operational Excellence: A critical component of organizational strategy focused on consistent and reliable execution of business activities, leading to superior performance in terms of cost, speed, and quality compared to competitors.

Operational Strategy: The specific actions and pathways an organization undertakes to effectively utilize its resources and achieve its operational goals.

Point Solution: A software or system designed to address a specific issue or manage a discrete part of a business process. Point solutions are limited in scope and integration, often requiring further adaptation to fit into broader enterprise systems seamlessly.

Process Reimagination: The holistic rethinking of business processes to improve efficiency, effectiveness, and adaptability by leveraging digital technologies and innovative approaches.

Prompt Engineering: The practice of designing and refining inputs (prompts) to artificial intelligence systems, especially in natural language processing, to elicit the most accurate and relevant outputs.

Prompting: The act of providing structured input to an AI system to trigger a specific type of response or action, crucial in interactions with models like chatbots or language models.

Quantum Computing: A type of computing based on the principles of quantum theory, which explains the nature and behavior of energy and matter on the quantum (atomic and subatomic) level. Instead of binary bits based on either 0 or 1 used for computing by typical computers as we know them today, Quantum Computer are based on qubits which are particle that are entangled and in a state of superposition. Quantum computers are capable of processing complex data at speeds unattainable by traditional computers.

Robotic Process Automation (RPA): A type of Artificial Intelligence (AI) used to automate any high-volume, stable, repeatable, and rules-based process across the applications utilized by people to carry out those processes with little need of human judgement.

Software Robot (or bot): A software algorithm programmed to perform automated, repetitive, pre-defined tasks. These bots can mimic human actions to interact with digital systems and perform tasks such as data entry, processing transactions, or managing workflows.

Solution Design (also called the "As-Is" Process): The process of defining the architecture, components, interfaces, and data flow of an Intelligent Automation solution tailored to meet specific business needs.

Strategic Initiative: Essential actions or strategic projects deemed crucial for securing an organization's long-term success and adaptability in a rapidly evolving market environment.

Straight-Through Processing (STP): Straight-Through Processing in the context of Digital Transformation and intelligent automation refers to the automated, uninterrupted execution of transactions from start to finish without manual intervention. This concept is integral to enhancing operational efficiency and accuracy across various business processes.

System Solution: A comprehensive organizational approach to resolving business challenges that integrates multiple functionalities across an organization to streamline processes and enhance overall operational efficiency. Unlike point solutions, system solutions are designed to be scalable and adaptable to various business needs.

To-Be Process: The future state of a business process as envisioned after the implementation of Intelligent Automation solutions, focusing on improved efficiency, effectiveness, and flexibility and is an integral part of the Solution Design process.

Transformer: A type of neural network architecture that relies on self-attention mechanisms to process input data in parallel and capture complex dependencies. Widely used in Generative AI, Transformers excel in handling sequences, such as text or time-series data, making them effective for tasks like natural language processing, image generation, and more. They enable models to generate coherent and contextually relevant outputs based on learned patterns from large datasets.

Virtual Reality (VR): A simulated experience that can be similar to or completely different from the real world, created by computer technology. It immerses users in a virtual environment, typically through the use of special headsets with screens and motion-tracking sensors and hand controllers or other input devi

References

Accenture. (2022). The Rise of Generative AI: Unlocking New Possibilities for Businesses. Accenture.

Ajay Agrawal, J. G. (2018). Prediction Machines, Economics of Artificial Intelligence. Cambridge, MA: Harvard Business Review Press.

Ajay Agrawal, J. G. (2022). Power and Prediction. In J. G. Ajay Agrawal, Power and Prediction. Harvard Business Review Press.

Anthropic. (n.d.). Claude. Retrieved from Claude 3: https://claude.ai/chats

Bain & Company. (2023). Unleashing the Power of Generative AI in the Enterprise. Bain & Company.

Capgemini. (2021). Intelligent Automation: Driving Digital Transformation for the New Normal. Capgemini.

Chatbot, O. A. (n.d.). Revolutionising Lead Generation & Customer Service: Offernet's AskFlowe Chatbot. Retrieved from https://www.bastionflowe.com/offernet-askflowe-chatbot-lead-generation-customer-service/.

Coombs, C. (2020). Will COVID-19 be the tipping point for the Intelligent Automation of work? A review of the debate and implications for research. International Journal of Information Management, 1-4.

Daugherty, P. R. (2018). Human + Machine: Reimagining Work in the Age of AI. In P. R. Daugherty, Human + Machine: Reimagining Work in the Age of AI. Cambridge, MA: Harvard Business Review Press.

DHL. (n.d.). Bring on the bots! How automation can break through bottlenecks. Retrieved from dhl.com: https://www.dhl.com/global-en/delivered/digitalization/logistics-automation.html

FasterCapital. (n.d.). How Employers Can Encourage And Support Upskilling Among Employees. Retrieved from FasterCapital: https://fastercapital.com/topics/how-employers-can-encourage-and-support-upskilling-among-employees.html

Forrester. (2022). The Future of Robotic Process Automation. Forrester.

Gartner. (2022). Hype Cycle for Artificial Intelligence. Gartner.

Gerald, K. N. (2019). The Technology Fallacy. In K. N. Gerald, The Technology Fallacy. Cambridge: MIT .

Google. (n.d.). Gemini Advanced. Retrieved from Gemini Advanced: https://gemini.google.com/app

Google. (n.d.). Google Machine Learning Education. Retrieved from Prompt Engineering for Generative AI : https://developers.google.com/machine-learning/resources/prompt-eng

Hardy, J. (2022, 11 22). The What and Why of Digital Transformation Strategy and the Future of Business. Retrieved from https://cxotechmagazine.com/: https://cxotechmagazine.com/the-what-and-why-of-digital-transformation-strategy-and-the-future-of-business/

Hardy, J. (2023, 11 25). The Importance of having an Enterprise Automation Development Mindset for Intelligent Automation. Retrieved from https://www.linkedin.com/pulse/importance-having-enterprise-automation-development-jonathan-t--60anc/?trackingId=yXzU%2BbH8QUOWoN2qaa51Lw%3D%3D: https://www.linkedin.com/pulse/importance-having-enterprise-automation-development-jonathan-t--60anc/?trackingId=yXzU%2BbH8QUOWoN2qaa51Lw%3D%3D

Harvard Business Review. (2023). How Generative AI Will Transform Business. Harvard Business Review.

IBM. (2022). Generative AI: Powering the Next Wave of Intelligent Automation. IBM.

(2022). Intelligent Automation: Unlocking the Full Potential of Digital Transformation. Deloitte.

IntellPaat. (n.d.). What Is Blue Prism: Architecture, Components, Features. Retrieved from https://intellipaat.com/blog/what-is-blue-prism/

KPMG. (2022). The Transformative Power of Intelligent Automation. KPMG.

Lucidchart. (n.d.). Retrieved from Lucidchart: https://www.lucidchart.com/blog/introducing-process-diagrams-from-csv-import

Marco Iansiti, K. R. (2020). Competing in the Age of AI: Strategy and Leadership When Algorithms and Networks Run the World. In K. R. Marco Iansiti, Competing in the Age of AI: Strategy and Leadership When Algorithms and Networks Run the World. Cambridge, MA: Harvard Business Review Press.

Marco Iansiti, K. R. (2020). Competing in the Age of AI: Strategy and Leadership When Algorithms and Networks Run the World. In K. R. Marco Iansiti, Competing in the Age of AI: Strategy and Leadership When Algorithms and Networks Run the World. Cambridge: Gildan Media.

McKinsey & Company. (2022). The Next Frontier of AI: Generative AI. McKinsey & Company.

OpenAI. (n.d.). Chat GPT-4. Retrieved from https://chat.openai.com/

P&G. (n.d.). Digital — The Secret to Solving Consumer Problems. Retrieved from http://us.pg.com: https://us.pg.com/blogs/investor-day-2022-series-digital-secret-to-solving-consumer-problems/

Peter Foy. (2023). Prompt Engineering: Advanced Techniques. Retrieved from MLQ.ai: https://www.mlq.ai/prompt-engineering-advanced-techniques/

Process mapping software. (n.d.). Retrieved from lucidchart.com: https://www.lucidchart.com/pages/examples/process-mapping-software

Process mapping software. (n.d.). Retrieved from www.lucidchart.com: https://www.lucidchart.com/pages/examples/process-mapping-software

Putting digital at the heart of strategy, When everyone is digital, strategy is the differentiator. (2021, April 22). Retrieved from Deloitte Insights: https://www2.deloitte.com/us/en/insights/topics/digital-transformation/digital-acceleration-in-a-changing-world.html

PwC. (2021). Robotic Process Automation in Insurance: Driving Efficiency and Accuracy.

QuantHub. (n.d.). How Data is Changing the Finance Industr. Retrieved from https://www.quanthub.com/how-is-data-changing-the-finance-industry/.

Ragu Gurumurthy, Jonathan Camhi, David Schatsky. (2020, May 26). Uncovering the connection between digital maturity and financial performance. Retrieved from Deloitte Insights: https://www2.deloitte.com/us/en/insights/topics/digital-transformation/digital-transformation-survey.html

Richard Horton, Justin Watson, David Wright. (2019, September 06). Automation with intelligence. Retrieved from Deloitte Insights: https://www2.deloitte.com/us/en/insights/focus/technology-and-the-future-of-work/intelligent-automation-technologies-strategies.html

Saldanha, T. (2019). Why Digital Transformations Fail: The Surprising Disciplines of How to Take Off and Stay Ahead. In R. A. Tony Saldanha, Why Digital Transformations Fail: The Surprising Disciplines of How to Take Off and Stay Ahead. Berrett-Koehler Publishers.

Technology Magazine. (2023, April 5). Cloud transformation helping airline group soar to its goals. Retrieved from Technology Magazine.com: https://technologymagazine.com/company-reports/lufthansas-cloud-journey-is-helping-group-soar-to-its-goals

Tim Smith, Tim Bottke, Gregory Dost, Diana Kearns-Manolators. (2023, January 31). Unleashing value from digital transformation: Paths and pitfalls. Retrieved from Deloitte Insights: https://www2.deloitte.com/us/en/insights/topics/digital-transformation/digital-transformation-value-roi.html

UiPath. (n.d.). AI-powered automation transforms operations for Omega Healthcare. Retrieved from UiPath.com: https://www.uipath.com/resources/automation-case-studies/omega-healthcare-boosts-efficiency-through-automation

UiPath. (n.d.). Efficiency Without Limits: UiPath and EY's Ambitious RPA Implementation. Retrieved from UiPath: https://www.uipath.com/resources/automation-case-studies/ey

UiPath. (n.d.). The Leading Global Corporate Immigration Firm Ups Its Automation Commitmen. Retrieved from uipath.com: https://www.uipath.com/resources/automation-case-studies/bal

UiPath. (n.d.). Vodafone Turkey inspires a new perspective on work. Retrieved from UiPath.com: https://www.uipath.com/resources/automation-case-studies/rpa-has-revolutionized-vodafone-turkey

Understanding Digital Transformation: Benefits, Examples, and Strategies. (2022, August 20). Retrieved from Enreach.com: https://enreach.com/en/news-knowledge/blog/understanding-digital-transformation-benefits-examples-and-strategies

Unknown. (2024, March 24). The Evolution of RPA: A 30-Year Journey. Retrieved from ElectorNeek: https://electroneek.com/rpa/history-of-rpa/

Unknown. (n.d.). Digital Sonar. Retrieved from Digital Sonar: https://www.digitalsonar.ro/expertise/digital-transformation

Vikram Mahidhar. (2014, January 22). Deloitte Insights. Retrieved from Intelligent automation: A new era of innovation: https://www2.deloitte.com/us/en/insights/focus/signals-for-strategists/intelligent-automation-a-new-era-of-innovation.html

Vittorio Cretella. (2022, December 5). https://us.pg.com/. Retrieved from Digital — The Secret to Solving Consumer Problems: https://us.pg.com/blogs/investor-day-2022-series-digital-secret-to-solving-consumer-problems/